Reptiles and Amphibians of Fort Ord Natural Reserve

Written by Max L. Taus

Illustrations by Sasha Taus, Stacy Wu, Charlotte Grenier, Tina Milz, Hannah Caisse, Mariam Moazed, Sofia Vermeulen, Elexis Padron, Willow Mosely, Dillyn Adamo, Grace Ackles, and Eleni Christoforou

Photographs by Max Taus and Gary Nafis

Reptiles and Amphibians of the Fort Ord Natural Reserve
By Max Taus

Illustrated by Sasha Taus, Stacy Wu, Charlotte Grenier, Tina Milz, Hannah Caisse, Mariam Moazed, Sofia Vermeulen, Elexis Padron, Willow Mosely, Dillyn Adamo, Grace Ackles, and Eleni Christoforou

With photographs by Max Taus and Gary Nafis

ISBN #9781986569347

Copyright © 2018 by Max Taus

Reptiles and Amphibians of the Fort Ord Natural Reserve is the property of Max Taus. With the exception of very brief extracts used as part of a book review, no part of this book may be reproduced or transmitted in any form or by any means, electronic, mechanical, photocopying, recording or otherwise, without prior written permission of the author. For permission, contact the author via maxlevitaus@gmail.com.

Designed, formatted and edited by Max Taus and Chris Lay

Published by UCSC Natural Reserves as a resource for students and community members of University of California, Santa Cruz

Notice of Rights

All photos in the guide were taken by the author except as otherwise noted. Original work of others that appears in this book is used with their permission and is protected by copyright, retained by the originators or their assignees. All rights reserved.

Special Thanks

The Ken Norris Center for Natural History at UCSC Student Project Award Program

Norris Center and Natural Reserve staff Chris Lay, Gage Dayton, Alex Jones,

Joe Miller, and Alex Krohn

Fort Ord interns Caleb Perez and Brandon Cluff

The author in the process of catching a Chuckwalla.

Contents

Reptiles and Amphibians of Fort Ord 1
Preface 2
Introduction 3
    Methods 4
    Plant Communities at Fort Ord 5
    Thermoregulation 7
    Foraging Strategies 8
    Reproduction 9
    Anti-predation Adaptations 10
    Conservation Note 12
    Dichotomous Keys 13
Species Accounts 18
    Lizards 19
    Snakes 41
    Turtles 63
    Frogs 65
    Salamanders 69
Acknowledgments 87
Glossary 88
Works Cited 95
Index 98
About the Author 99

Reptiles and Amphibians of the Fort Ord Natural Reserve

Amphibians: Ambystomatidae
Ambystoma californiense — California Tiger Salamander

Amphibians: Plethodontidae
Aneides lugubris — Arboreal Salamander
Batrachoseps gavilanensis — Gabilan Slender Salamander
Ensatina eschscholtzii eschscholtzii — Monterey Ensatina

Amphibians: Hylidae
Pseudacris regilla — Pacific Chorus Frog

Reptiles: Anguidae
Elgaria coerulea — Northern Alligator Lizard
Elgaria multicarinata — Southern Alligator Lizard

Reptiles: Anniellidae
Anniella pulchra — Northern California Legless Lizard

Reptiles: Phrynosomatidae
Phrynosoma blainvilli — Coast Horned Lizard
Sceloporus occidentalis — Western Fence Lizard

Reptiles: Scincidae
Plestiodon skiltonianus skiltonianus — Skilton's Skink

Reptiles: Colubridae
Coluber constrictor mormon — Western Yellow-bellied Racer
Contia tenuis — Common Sharp-tailed Snake
Diadophis punctatus vandenburghii — Monterey Ring-necked Snake
Lampropeltis getula — California Kingsnake
Pituophis catenifer catenifer — Pacific Gopher Snake
Thamnophis elegans terrestris — Coast Garter Snake

Reptiles: Viperidae
Crotalus oreganus oreganus — Northern Pacific Rattlesnake

Reptiles: Emydidae
Actinemys marmorata — Western Pond Turtle

Preface

I've always been more intrigued than intimidated by reptiles and amphibians. So many of them have peculiar morphologies and ecologies. Ever since I was a child, I dug up slender salamanders in my backyard, hunted for tadpoles while learning to swim, and caught fence lizards basking on the sidewalk. My innate curiosity drove me to try and catch everything I could get my hands on. In college, it drove me to study herpetology. I quickly became involved with the naturalist community in Santa Cruz and decided to double major in Environmental Studies and Molecular Biology. I participated in a ten week field studies class, Natural History Field Quarter, which took place all over California at various UC Natural Reserves. This deepened my love of the natural world and stirred a desire to work on the UC Natural Reserves. I later helped monitor the population of the endangered San Francisco Garter Snake at the Año Nuevo reserve, which transitioned to my fieldwork at Fort Ord Natural Reserve (FONR) in 2017. I was thrilled to have the opportunity to produce this guide for future herpetology work at one of UCSC's Natural Reserves.

From January through May of 2017, I conducted weekly surveys throughout the winter and spring. Four amphibian species and eleven reptile species were found: one chorus frog, three plethodontid salamanders, five lizards and six snakes. Of the nineteen species whose ranges are thought to extend into the greater Marina area, only one species (Northern Alligator Lizard) has not been observed on the FONR. The abundant species diversity of the reserve reflects the wide array of habitat types found among the ancient inland coastal sand dunes of the Monterey Bay.

Sexually mature adult male Skilton's Skink
llustration by Grace Ackles.

Introduction

In 1991, the United States Army announced the closure of Fort Ord, the largest military base in California, located along the Monterey Bay. When the Army left in 1994, ideas for how to use this newly available land were discussed at length, with ideas ranging from building a second Disneyland to redistributing land for agriculture and housing projects that could triple the population of Seaside. Two-thirds of the base's 27,879 acres was undeveloped and was used almost exclusively as an artillery and munitions range. In June 1996, the UC Natural Reserve System received a 605-acre plot of Fort Ord's land adjacent to the airport, which was never used as a munitions or bombing range. This parcel is comprised of near-pristine maritime chaparral habitat. Since its inception, FONR has been used by an increasing number of classes and researchers in the local Monterey Bay area. In addition to many other rare species, the reserve hosts three sensitive reptile and amphibian species: California Tiger Salamander, California Legless Lizard, and the Coast Horned Lizard. A number of these species' populations have locally declined due to habitat loss and fragmentation via development, the proliferation of invasive ants displacing native food sources, illegal pet trade activities, and increased predation from domestic dogs and cats.

The Roots of Herpetology

Our use of the word "herp" to refer to reptiles and amphibians originally derived from the Greek word "herpein", which means to creep or crawl. Although reptiles and amphibians have been classified together because they share this type of locomotion, each group has developed from separate evolutionarily distinct lineages. **Reptiles** are vertebrate animals with dry-scaly skin leading mostly terrestrial lives and (typically) laying soft-shelled eggs on land. **Amphibians** are distinguished by having moist, permeable skin. Many lead an aquatic gill-breathing larval stage followed by a terrestrial lung-breathing adult stage. Herpetologists study the ecological, behavioral, morphological, and physiological characteristics of both reptiles and amphibians.

Methods

Over the summer of 2016, I placed forty 4' x 4' x 0.75" plywood coverboards in various habitat zones across the northern and southern sections of the reserve. I later added a 41st coverboard in the southern section that had been previously used by small mammals. Eleven weeks of six-hour surveys were performed from January to May of 2017, on warm sunny days from 10am to 4pm. Survey methods included peering into gopher holes and tree caverns, digging through oak duff and loose soil and turning over ground cover such as logs, rocks, and each of the coverboards. All specimens I found were caught by hand (when possible), identified, photographed, sexed, any peculiarities in behavior or appearance noted and their location recorded in an ArcGIS database. Joe Miller (current Field Manager at FONR) and I established an ArcGIS database for the purpose of consolidating all observations made on the reserve for future research.

Looking under a coverboard placed in grassland habitat.

Major Plant Communities at FONR

The reptiles and amphibians of the Fort Ord Natural Reserve occur in four major habitat-types: Coast Live Oak Woodland, California Annual Grassland, Coastal Scrub, and Maritime Chaparral. These four habitat types dominate the FONR landscape, with each habitat characterized by it's own unique plant diversity.

Maritime Chaparral

FONR soils are mainly composed of inland dunes of sand blown in from the Monterey Bay coastline. With few available nutrients and a low water holding capacity, these soils support a diverse number of evergreen shrubs, including: Sandmat Manzanita (*Arcostaphylos pumila*), Woollyleaf Manzanita (*Arctostaphylos tomentosa*), Monterey Ceanothus (*Ceanothus rigidus*) and Mock Heather (*Ericameria ericoides*). Other commonly associated shrubs include Chamise, Toyon, and California Poison Oak. The coastal fog influence makes this chaparral community unique from most of the more inland chaparral communities found throughout California.

Coast Live Oak Woodland

The overstory of this community is dominated by short but long-lived Coast Live Oaks (*Quercus agrifolia*). This is likely due to the relatively dry soils and a longer fire return interval than many other chaparral dominated habitats. The understory consists of shade tolerant shrubs such as Poison Oak and California blackberry, and herbaceous plants such as Bracken Fern, Fiesta Flower, Miner's Lettuce and Bermuda Buttercup (an invasive). In drier areas where oaks are more widely-spaced, the understory may consist almost entirely of exotic annual grass species with few shrubs. Where the habitat intergrades with coastal scrub, typical understory species are Bush Monkeyflower, Coyote Brush, Black Sage, and California sagebrush. With the absence of significant fires on the reserve in the past few decades, the Coast Live Oak portions of the reserve have been encroaching on the maritime chaparral habitats and consequently, many of the low-growing Manzanitas are being crowded out as a result.

California Annual Grassland/ Degraded Coastal Prairie

Introduced annual grasses are the dominant plant species in this habitat type. These invasive grasses include Wild Oats (*Avena spp.*), Soft Chess (*Bromus hordeaceus*), Ripgut Brome (*Bromus diandrus*), and Red Brome (*Bromus madritensis rubens*). Some native perennial grasses mix in with the dominant annuals, including Purple Needlegrass (*Stipa pulchra*). The common forbs include Broadleaf Filaree, Redstem Filaree, Turkey Mullein, True Clovers, Bur Clover, Popcorn Flower, and many others. The California Poppy (*Eschscholzia californica*) and Sky Lupine (*Lupinus nanus*) are ubiquitous in this habitat.

Coastal Scrub

The common overstory species in Coastal Scrub at FONR are Bush Monkeyflower, California Blackberry, Poison Oak, Black Sage, California Sagebrush, and Coffeeberry. Understory clusters of Owl Clover, Sand Gilia, Monterey Spineflower, Buckwheat, Tidytips, and Broom Rose are all dispersed throughout the reserve. Patches of open dune habitat, where few if any plants grow, are dispersed throughout the reserve in each of the four main plant communities (especially Maritime Chaparral and Coastal Scrub). This open habitat is expecially vital in order for reptiles to effectively thermoregulate.

Relatively open dune habitat (foreground); Sandmat Manzanita (middle) and Oak Woodland (background).

Natural History of Reptiles and Amphibians

Thermoregulation

There are two ways that organisms regulate and maintain their core temperature within a certain range: ectothermy (cold-blooded) and endothermy (warm-blooded). The main difference between these two types of thermoregulation is how each animal reaches and maintains its core operating temperature. Ectotherms rely on external sources of heat, while endotherms rely on internal heat production. Reptiles and amphibians are both ectotherms and have adapted to **behaviorally thermoregulate** by moving in and out of sunlight and using their surrounding environment to heat and cool their core body temperature. By absorbing solar radiation directly, conducting heat by basking on warmer or cooler substrates and seeking shelter in nearby shrubs or holes, reptiles and amphibians can maintain their optimum operating temperature. Amphibians rely heavily on bodies of water to stay moist year-round and keep from overheating, while reptiles manage to acquire most of their moisture from their prey.

Sit-and-wait predators easily shift their core body temperature by moving from a sunny to a shadier perch, while widely foraging predators struggle to thermoregulate, because they're constantly being exposed to solar radiation while continuously searching for prey. Body size also dictates how effectively reptiles and amphibians can thermoregulate. Larger organisms cool down slowly and can spend more time in a shaded environment, in contrast with smaller organisms which must constantly shift in and out of the sunlight to maintain their core temperature. In addition to shifting from cooler to warmer substrates, lizards can also alter their skin color and body conformation to facilitate heat absorption and loss.

Coast Horned Lizards use their large ventral surfaces to help thermoregulate.

Foraging Strategies

Reptiles and amphibians utilize one of three types of foraging strategies: sit-and-wait, widely foraging, and cruising. As their name suggests, sit-and-wait predators spend a vast majority of their time waiting for prey to come close enough to quickly snatch up. Widely foraging predators are constantly roaming in search of their next meal. Cruising foragers are intermediates between both extremes and are moderately active predators. An organism's foraging strategy directly influences the morphology and ecology of each organism. Sit-and-wait predators like the Western Fence Lizard and Coast Horned Lizard typically have a stocky build, are well camouflaged, and heavily rely on vision for discerning large mobile prey. Widely foraging predators like alligator lizards have elongated bodies with limited vision and a developed sense of smell to track down prey. Exceptions to these categories include rattlesnakes, which are sit-and-wait predators with relatively poor vision. These snakes have a well developed Jacobson's organ to perceive olfactory cues and an injectible venom system to compensate for their lack of mobility. Both reptiles and amphibians use their Jacobson's organ to detect minute particle gradients in the atmosphere. This allows them to track prey and sense mates for reproduction.

Risk of predation for sit-and-wait predators is relatively low due to their camouflaged nature, while widely foraging predators are consistently preyed upon at a higher rate. Sit-and-wait predators occupy a relatively small area and are often territorial; in contrast, widely foraging predators cover a large area and are less territorial due to their meandering nature. Widely foraging predators tend to consume smaller sedentary prey found while foraging through duff or under rotting logs, while sit-and-wait predators tend to prey on larger, widely foraging predators traversing active trails. Due to widely foraging predators' mobile nature, they typically carry a smaller clutch of eggs to compensate for the hindrance to mobility associated with being gravid. In contrast, sit-and-wait predators can afford to carry a larger cluster of eggs because of their sedentary behavior.

Reproduction

Courtship between reptiles and amphibians has evolved to incorporate territorial displays and chemical communication using pheromones. Male reptiles and amphibians often defend their territory and display their dominance by performing a series of head nods and "pushups" or by fighting with other males that wander too closely. All reptiles have internal fertilization (zygote formation occurs inside the body), while many amphibians reproduce via external fertilization (zygote formation occurs outside the body). Three reproductive modes exist among reptiles and amphibians: oviparity (egg-laying), viviparity (live-bearing), and an intermediate form known as ovoviviparity where eggs are fertilized internally and remain within a female's body until they are ready to hatch. Viviparous species typically produce fewer and larger individuals while oviparous species produce many smaller offspring. Parental care is uncommon but has been observed in Western Skinks and Monterey Ensatinas, which will stay and guard their eggs until they hatch. Viviparity has evolved primarily in colder climates, where controlling an embryo's internal development is the most crucial aspect of reproduction.

During copulation, male reptiles will typically bite hold of the female's neck, evert their hemipenes, (which appear like two pairs of bifurcated penises) and insert one into the female's cloaca. Many male Fence and Horned Lizards contain special glands in their hindlegs that exude secretions during copulation that increase a female's receptiveness towards the male. When a female is ready to mate, she'll allow a courting male to mount her in amplexus (where pheromone exchange occurs) and press his cloacal region against hers (where sperm exchange occurs). In amphibian amplexus, a male will either fertilize a female's clutch externally once the eggs have already been lain or will deposit a spermatophore on the ground, which the female will pick up with her vent to fertilize her eggs. Many reptiles and amphibians bury their eggs in moist soil, underneath rotting logs or duff, while others lay their clutches together in or near bodies of water, typically adjacent to the organism's primary food source.

Anti-Predation Adaptations

Caudal autotomy is a common anti-predation adaptation developed among some lizards and salamanders that allows them to intentionally detach their tail to escape predation. Because of this, the standard method of measuring reptiles and amphibians is to ignore tail length and measure from snout (head)-to-vent (cloaca). Because lizards and salamanders rely on their tails for fat storage, balance, and mating displays, they each have varying propensities to perform caudal autotomy. Shedding a tail can be more harmful than helpful at times, especially for widely foraging predators who rely on their tail for balance while chasing prey or avoiding other predators. Once lost, tails can be regrown but not without costs to the organism. In lizards, tails break off at weak points located along the tail's vertebrae. These lost vertebrae are replaced with a cartilaginous substitute when the tail regenerates. As a result, subsequent tail breaks occur proximal to the regenerated tail portion. Salamanders capable of tail regeneration also break their tails off at weakened points along their tail vertebrae. However, their regenerative capabilities far exceed that of lizards. In addition to regrowing their tails, salamanders are able to regenerate entire limb segments and parts of damaged vital organs!

Lizards and salamanders can either drop their tails during or before being attacked by a predator. Their shed tails contain special nerve endings that cause the tail to wildly flail about for several minutes after it is released. This is a very effective distraction technique to avoid being eaten. Certain lizards and salamanders can also control how much of their tail is shed, depending on various external influences. For instance, if a lizard or salamander is well fed or in colder climates, they are more likely to drop a larger segment of their tail (since they rely less on their tail for energy storage) to give themselves more time to escape.

Another common anti-predation adaptation practiced by various reptiles is known as musking. Most commonly practiced among snakes, musking involves excreting a foul-smelling substance produced in the cloaca to deter natural predators. In addition, some species use musking to communicate with the opposite sex.

Studies have shown that gravid female snakes, which are less able to escape via flight, are more reliant on musking than male and non-gravid females. These female snakes will either produce more musk or a more repulsive musk to compensate for their lack of mobility. Although lizards don't produce musk, when disturbed, a number of lizards (such as Alligator or Legless Lizards) will readily smear their cloacal contents onto any potential predators. For obvious reasons, this smearing of cloacal contents has also proven effective as a predator deterrent.

Horned Lizard photo taken after performing ocular autohemmorhaging. Taken from https://amazingadaptations.weebly.com/texas-horned-lizard.html

Southern California Legless Lizard that recently performed caudal autotomy.
Photo taken by Gary Nafis from http://californiaherps.com/lizards/images/apulchrarc411.jpg

Conservation Note

Reptile and amphibian populations have declined globally over the past few decades. These declines are likely from habitat fragmentation and loss, pesticide and fertilizer runoff, novel diseases and introduced invasive species. In addition, human exploitation in the pet trade of specific reptile and amphibian species has also led to dramatic population declines.

Though often take for granted, reptiles and amphibians provide direct ecological benefits for humanity. Many snakes maintain rodent populations, which act as vectors for disease. Western Fence Lizards cure ticks that feed on them of Lyme disease. Proteins found in Gila Monster venom have been used to treat diabetes and breast cancer. Russell's Viper venom may contribute to a potential cure for Alzheimer's disease. And the Mexican Salamander's (Axolotl) ability to regenerate its limbs motivates research in hopes of reverse engineering this process in people.

Even with all these wonderful ecological and medicinal benefits, many are still disturbed by reptiles and amphibians. Fear of snakes, for instance, is thought to affect as many as one in three adults worldwide. This phobia likely stems from our innate fear of the unknown and the historical references that associate snakes with evil. Physiologists at the Karolinska Institute and Hospital in Sweden have even shown that humans likely evolved to inherently interpret reptiles as a threat. As a result, reptiles are often killed out of fear, especially venomous snakes, which are rounded up and killed in the thousands annually in some parts of the USA.

This primal fear humans may carry can be lightened by acknowledging the inherent benefits that reptiles and amphibians provide to society and by spending time observing and appreciating them in their natural environment. And once we make a personal connection with these beautiful creatures, I believe it's possible to short-circuit our instinctual fears and give these animals the respect, protection and admiration they deserve.

Dichotomous Keys

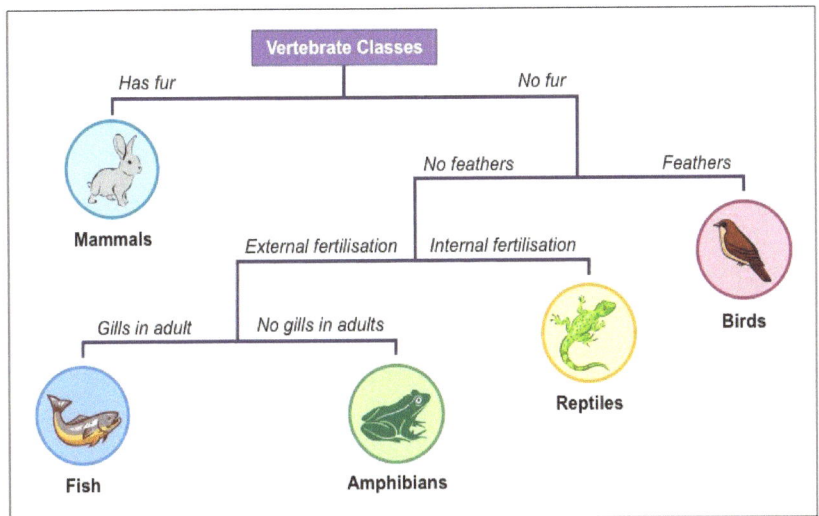

An example of a dichotomous key; used to group together organisms into clades, based off the presence or absence of key features. Image taken from http://ib.bioninja.com.au/_Media/dichotomous5_med.jpeg

A convenient way to relate different organisms is to group them into species, which by one common definition are groups of interbreeding organisms that are reproductively isolated from other such groups. Taxonomists categorize groups of species based on our understanding of evolutionary history as well as comparing each species' morphological and physiological traits. Certain definitive characteristics that have evolved over the millennia are used as major distinctions between related species groups. Below are dichotomous keys that can be used to identify the salamanders, lizards, and snakes at FONR. Since there is only one species of turtle and frog found on the reserve, they are not included as part of the key. Use the dichotomous keys, by choosing one of two statements that best matches the description of the specimen in question. Each stement will either lead to an ID or further descriptions, eventually leading to an ID.

Key to Salamanders

Note: California Salamander species have four legs and a tail, and may at first appear to resemble some lizard species. However, they lack scales and claws and typically have moist skin.

1a. Nasolabial groove between nostril and lip present; skin smooth, never rough/granular (Lungless Salamanders/Plethodontidae) .. **3a**
1b. Nasolabial groove absent; skin may be smooth or rough **2a**
2a. Skin rough, except in males during breeding season; dorsal color various shades of solid brown; ventral color solid yellow to orange .... **Newts/*Taricha* (Not at FONR)**
2b. Skin smooth; dorsal coloration mottled or spotted .... ***Ambystoma*/ California Tiger Salamander**
3a. Base of tail constricted .... ***Ensatina*/ Monterey Ensatina**
3b. Tail not constricted at the base **4a**
4a. Four toes on all four feet; very small legs on a long, wormlike body .... ***Batrachoseps*/ Gabilan Slender Salamander**
4b. Four toes on front feet, five on hind feet; not wormlike in appearance **5a**
5a. Adult males usually have projecting upper jaw teeth that can be felt by stroking upward with your finger; toe tips more square than rounded; no distinct stripe on dorsal side .... ***Aneides*/Arboreal Salamander**
5b. Upper jaw teeth usually not detectable by stroking upward; toe tips round, not squared off; dorsal stripe usually present .... **Woodland Salamanders (Not at FONR)**

A Young Monterey Ensatina strolling around grassy soil.

Key to Snakes

Note: All snakes lack limbs and moveable eyelids (legless lizards have moveable eyelids).

1a. Tail with a rattle or single "button" at tip; two pairs of small openings on snout (nostrils and loreal pits) .... *Crotalus*/ **Northern Pacific Rattlesnake**

1b. Tail without rattle and snout without paired pits **2a**

2a. Dorsal scales keeled (ridged down the middle) **3a**

2b. Dorsal scales smooth (not ridged down the middle) **4a**

3a. Usually four prefrontal scales in front of eyes; large brown blotches on a tan-yellow background; eye pupils round, not vertical .... *Pituophis*/ **Pacific Gopher Snake**

3b. Two prefrontal scales in front of eyes; bold to faint mid-dorsal stripe usually present; single anal scale anterior to vent .... *Thamnophis*/ **Coast Garter Snake**

4a. Adults with olive-green dorsal and pale yellow ventral color; juveniles with brown blotches on tan background, similar in appearance to Gopher Snake, but with two prefrontal scales and noticeably large eyes ..... *Coluber*/ **Western Yellow-bellied Racer**

4b. Adult coloration not as in 4a **5a**

5a. Sharp spine-like point at tip of tail; all ventral body scales with regular narrow black crossbands .... *Contia*/ **Common Sharp-tailed Snake**

5b. No spine-like point at tip of tail; ventral scales usually not marked with black crossbands **6a**

6a. Plain dorsal coloration without pattern; head often darker; narrow orange or white dorsal neck band usually present .... *Diadophis*/ **Monterey Ring-necked Snake**

6b. Dorsal color pattern of spots, blotches or crossbands **7a**

7a. Bold (black & white) dorsal crossbands that extend across ventral body .... *Lampropeltis*/ **California Kingsnake**

7b. Dorsal crossbands or brown and tan mottled pattern present, but without bold banding across entire ventral surface .... **Not found on reserve**

Key to Lizards

<u>Note</u>: All lizards have scaled skin, legged forms have clawed toes, and most species have moveable eyelids.

1a. Eyes without opaque eyelids . . . . See Snakes
1b. Moveable eyelids present 2a
2a. Legless, snakelike body; tiny eyes with moveable eyelids that blink . . . . *Aniella/* **Northern California Legless Lizard**
2b. Limbs present 3a
3a. Body (except for head) covered with smooth, shiny, cycloid (rounded edges) scales . . . .*Plestiodon/* **Skilton's Skink**
3b. Scales not cycloid (rounded edges) over entire body 4a
4a. Projected structures on back of head, ranging from spikes to small nubbins; usually one or two rows of enlarged, pointed scales along sides of the body . . . .*Phrynosoma/* **Coast Horned Lizard**
4b. No "horns" or enlarged, pointed lateral scale rows 5a
5a. Distinct lateral skin fold on body extending from fore to hind limbs, containing small granular scales on sides of body that separate large, squarish dorsal and ventral scales 6a
5b. No lateral skin folds separating dorsal and ventral scales 7a
6a. Dark marking in between ventral scales (prefer shadier/colder environments). . . . .*Elgaria/* **Northern Alligator Lizard**
6b. Dark marking down the middle of ventral scales (prefer sunnier/warmer environments). . . . .*Elgaria/* **Southern Alligator Lizard**
7a. All dorsal scales keeled and pointed; incomplete gular fold on throat skin . . . . *Sceloporus/* **Western Fence Lizard**
7b. Some or all dorsal scales granular; if keeled scales present, they are never pointed; complete gular fold on throat skin . . . .**Not found on reserve (Zebra-tailed, Fringe-toed, Collared, Leopard, Side-blotched, Brush, Tree, and Rock Lizards)**

Key to Turtles

<u>Note</u>: Physical features such as the shape and position of the epidermal shields of the carapace (shell) and limb anatomy are usually used to differentiate families, genera and species. However, in all of California, there is only one native species.

............................*Actinemys*/ **Western Pond Turtle**

Comparison of shells of native pond turtle and the non-native eared slider, which are not found at FONR but are sometimes found in human-altered freshwater ponds. Figure by Gary Nafis from http://www.californiaherps.com/turtles/images/sliderpondinguinals.jpg

Key to Frogs

<u>Note</u>: All frogs and toads have scaleless skin and distinctly larger hind legs than front legs. They also do not have a true tail.
.............................*Pseudacris*/ **Pacific Chorus Frog**

Green morph of a Pacific Chorus Frog.

Western Fence Lizard
Sceloporus occidentalis

Male Western Fence Lizard in "push-up" pose. Illustrated by Mariam Moazed.

Description

These lizards are 2.25 to 3.5 inches long from snout to vent. A fairly small lizard with uniformly sized, keeled scales on the back, sides and belly. Scales on the backs of the thighs are mostly keeled and abruptly smaller, with bright yellow to orange coloration. Dorsal color is brown, gray, or black with blotches, which may form irregular bands or stripes. The ventral side is light in color with a vivid blue belly and throat, hence the colloquial name "Bluebellies." Juveniles have little or no blue coloration on the throat and faint blue belly markings to none at all. Male fence lizards have blue markings on the sides of the belly edged in black, two blue patches on the throat, two enlarged post-anal scales, enlarged femoral pores and a swollen tail base. Females have less intense blue coloration on the belly, uniformly sized post-anal scales and dark bars on the back.

Behavior

Fence lizards are diurnal and often seen basking on rocks, downed logs, trees, fences, and walls on sunny days. They are most active when temperatures are warm. They become inactive during periods of extreme heat or cold during which they seek shelter in crevices, burrows, at the bases of bushes, underneath rocks, boards, tree bark and other debris.

Sometimes it's a scramble to catch all the specimens underneath a coverboard. This particular board was home to four fence lizards.

Fence lizards are prolific across the state of California, perhaps due to their success even in urban environments. Males establish and defend a territory with elevated perches, where they perform for receptive mates and fend off potential rivals. Males defend their territory and try to attract females with head-bobbing and a series of "push-ups" that exposes their brilliant blue throat and ventral colors. Territories are ultimately defended through physical confrontation between mature males. When disturbed, fence lizards can easily detach their tails to distract a potential predator, allowing them to escape. They eat small, mostly terrestrial, invertebrates such as crickets, spiders, ticks, and scorpions, and occasionally eat small lizards including smaller fence lizards! Females dig small holes in loose damp soil, where they'll lay clutches of 3 to 17 eggs from May to July.

Habitat

Fence lizards are found in a wide variety of habitat types including: Coast Live Oak woodlands, California Annual Grasslands, Coastal Scrub, and Maritime Chaparral.

Ventral side of Female vs Male Fence Lizard. Illustration by Grace Ackles.

Male Western Fence Lizard found basking on a rock.

Western Fence Lizard that performed caudal autotomy after being caught. Note its tail in the lower right of the picture.

Interesting Facts

In California, Western Black-legged Ticks, a.k.a. Deer Ticks, are the primary vectors for a bacteria (Borrelia burgdorferi) that causes Lyme disease. Deer Ticks often feed on fence lizards, which carry an unidentified protein speculated to bind with and kill the bacteria (Borrelia burgdorferi) found in the tick's gut. As a result, the presence of large fence lizard populations has been linked to tick populations with lower rates of Lyme disease!

Notes

Western Fence Lizards were by far the most prolific reptile I found on the reserve and were observed in every habitat type. I often found them being parasitized by nymphal ticks, located around the neck region or inside the earhole, attached in between scales. I most often observed them at the base of bushes, basking along frequently used paths and scurrying to safety to a nearby bush or gopher hole at the first sign of danger. Often observed with darker coloration during the start of the day and lighter by the end of the day due to melanophores helping facilitate heat absorption.

Skilton's Skink
Plestiodon skiltonianus (Eumeces skiltonianus)

Juvenile Skilton's Skink. Illustration by Grace Ackles.

Description:

This lizard measures 2.1 to 3.4 inches long from snout to vent. This small lizard has a smooth, shiny body with cycloid (rounded edges) scales, a small head, thick neck and small legs. Their dorsal side is striped with three dark brown and light cream stripes that extend along the body, while the ventral side is pale to gray. The tail of juveniles is bright blue, while older adults' tails are typically grayish. Younger adults will often retain some of their adolescent coloration. During the breeding season, sexually mature adult males will develop reddish-orange coloring on the face, chin, tail and occasionally, the sides of their body. This color acts as an invitation to receptive females.

Sexually mature adult male Skilton's Skink. Illustration by Grace Ackles.

Side by side comparison of adult female and male Skilton's Skink during the breeding season, on the left and right respectively.

Sexually mature adult male Skilton's Skink, displaying its reddish-brown coloration in the breeding season.

Behavior

Skinks are diurnal, but secretive and not typically seen when active. Occasionally, they are seen foraging in leaf litter or meandering through grasslands. They're most commonly found underneath bark, rocks and other debris, where they live in extensive burrows. Skinks will readily drop their tail, which writhes back and forth to distract a predator, while the lizard escapes. The bright blue coloring on the tail of a juvenile skink tends to distract a predator from the main body of the lizard. Skinks feed on insects and other small invertebrates, especially spiders and pill bugs. Females lay 2 - 10 eggs in June and July. A relatively uncommon trait these lizards possess is that females will stand guard and protect their eggs until they hatch. Eggs hatch in late July and August.

Habitat

These lizards prefer Open Grassland, Oak Woodland and Chaparral habitats on the reserve.

Interesting Facts

Skinks' brightly colored tail adaptations are suspected to have evolved as a reaction to high levels of avian predation; due to birds' keen sense of vision, the brightly colored tail acts as a beacon, directing strikes away from the body, increasing chances of survival.

Skilton's Skink caught while meandering through California Annual Grassland.

Notes

I observed an adult and juvenile Skilton's Skink occupying the same cover board in grassland habitat four weeks in a row. On the fifth week, I found they had been replaced with a California kingsnake, which I speculate ate one or both of the board's previous inhabitants. I also observed that a few of the juvenile skinks I had previously found in similar locations had lost their tails in the subsequent weeks and were in the process of regenerating new ones.

Coast Horned Lizard
Phrynosoma blainvilli

Coast Horned Lizard. Illustration by Charlotte Grenier.

Description
Adults are 2.5 to 4.5 inches long from snout to vent. A flat-bodied lizard with a wide oval-shaped body, enlarged pointed scales scattered along the upper body and tail and a large crown of horns along the head, with the two longest horns in the center. Males have two enlarged post-anal scales, enlarged femoral pores and a swollen tail base during the breeding season. Each side of the body has two rows of pointed fringed scales and each side of the throat has two or three rows of enlarged pointed scales. Dorsal coloration typically corresponds to the lizard's main habitat type, ranging from reddish-brown, to yellow or gray, with dark blotches on the back and large dark spots on the sides of the neck. The venter (belly) is cream, beige, or yellow, usually with dark spots and smooth belly scales.

Behavior
Horned lizards are diurnal and active during periods of warm weather, retreating underground and becoming inactive during extended periods of low temperatures or extreme heat. When threatened, they are capable of short bursts of speed in order to escape to thicker cover. Their primary defense is remaining motionless and relying on their camouflage. In some cases, Horned Lizards change their color slightly to better match their surroundings and will even shake their body from side to side to partially bury themselves in loose soil.

When threatened, this lizard inflates with air, making it larger and hard to swallow and opens its mouth and makes hissing noises as a warning. When grabbed, they will readily bite and vigorously squirm and jab their spiked body into an aggressor. As a last resort, they can squirt an aimed stream of blood from their eye towards the antagonizer. This excreted blood has been found to repel coyotes and foxes and possibly other predators. The Coast Horned Lizard eats mainly ants, especially harvester ants, but also consumes other small invertebrates such as spiders, beetles, termites, flies, honeybees, moth larvae and grasshoppers. It lays 6 to 21 eggs (averaging around 12) from May to June. Eggs hatch from August to September. Some females may lay two clutches of eggs in a year.

This female Coast Horned Lizard was seen in the Maritime Chaparral at the outskirts of a Sandmat Manzanita patch. After handling her for some time, she began to perform ocular auto-hemmorhaging, but instead of shooting an aimed stream of blood out from her eye, she shed a single tear of blood.

Habitat

Horned lizards inhabit open areas of sandy soil and low vegetation, often seeking refuge near the sandmat manzanita in the maritime chaparral and the California Sagebrush and Sticky Monkey Flower in the Coastal Scrub. They are frequently found posted near anthills.

Interesting Facts

Due to their popularity on the exotic pet trade, certain horned lizards have seen declines in population. The proliferation of non-native Argentine Ants has been particularly harsh on the Coast Horned Lizards on the reserve, which feed primarily on the native ants (especially Harvester Ants). Studies have shown that Coast Horned Lizards heavily rely on native ants for a healthy diet, but when fed a diet of primarily Argentine Ants, they begin losing weight and starving to death. The native Harvester Ants that are crucial for the survival of Coast Horned lizards are being outcompeted by invasive Argentine Ants on the FONR. Habitat fragmentation from agricultural and urban development in conjunction with a decline in native ant populations and an increase in domestic animal predation has led to drastic declines in Coast Horned Lizard populations. As of 2017, this species has been listed with the California Department of Fish and Game (CDFG) as a species of special concern and should be considered for protection under the endangered species act within the near future.

Coast Horned Lizard found in the Coastal Scrub, frequently observed hiding at the base of California Sagebrush and Sticky Monkey Flower.

Male Coast Horned Lizard observed rummaging through oak duff.

Notes

I often observed this lizard basking at the base of Sandmat Manzanita, California Sagebrush, Coyote Brush, and Bush Monkey Flower with the occasional sighting under the Coast Live Oaks. When confronted with danger, these lizards will often remain motionless, which has led to them being crushed by cars and agricultural machinery; during my surveying, I found a few horned lizards that were flattened (by cars) on the fire roads of the reserve. One Coast Horned Lizard that I captured shed a tear of blood from its eye, but apparently chose not to squirt a full stream of blood at me. I'd occasionally observe these lizards being parasitized by mites along the neck region. I have seen a few instances of Horned lizards that were attempting to feed from anthills, being swarmed by invasive Argentine ants, in the Maritime Chaparral. I would suggest a study to determine how the distribution of invasive Argentine Ants on the reserve is affecting Coast Horned Lizard distribution and populations on the FONR.

Alligator Lizards
Elgaria spp.

Large bony scales and a large head with an elongated body and powerful jaws (similar to alligators) give these lizards their common name. These lizards are characterized by a thick, rounded body, with short limbs and a long tail, which can often be twice the length of its body, if it has never been broken off and regenerated. Scales are keeled on the back, sides and legs, with a band of small granular scales separating the larger reinforced scales on the dorsal and ventral sides. This creates a distinct lateral fold along each side of its body. These folds allow the body to expand to hold food, eggs or live young and contract when the extra space isn't needed. Dorsal coloration is highly variable, typically correlating with the preferred habitat type of the lizard.

Alligator Lizards move with a snake-like undulating motion, often tucking the rear legs up against the side of the body and pulling along on their belly with their front two feet. They use their prehensile tail to climb trees in search of food and escape predation. The tail of an alligator lizard is easily broken off and will grow back. When first detached, the tail will writhe around for several minutes, long enough to distract a hungry predator from the lizard. Other defensive tactics used by alligator lizards are biting and smearing the contents of the cloaca on their enemy. They often bite onto the head of a predatory snake, rendering the snake unable to attack. When handled by people, they will readily bite and can draw blood. One observer reported seeing a juvenile alligator lizard bite onto its own tail, making itself impossible to be swallowed by a snake, which eventually gave up!

These ubiquitous lizards are an example of a classic feeding generalist, consuming a wide variety of prey, including invertebrates, lizards, bird eggs and small mammals. During the spring breeding season, a male lizard grabs on to the head of a female with his mouth until she is ready to let him mate with her. They can remain attached this way for many hours, almost oblivious to their surroundings. Besides keeping her from running off to mate with another male, this probably shows her how strong and suitable a mate he is.

Alligator Lizards that live in colder climates (like the Northern Alligator Lizard) are more often often observed giving live birth to allow for more parental control over the developing embryo's core temperature. In contrast, the Southern Alligator Lizard (which lives in warmer climates) lays eggs instead.

Northern and Southern Alligator Lizards are difficult to tell apart. For both species, the head of a male is broader than a female's with a more triangular shape. The iris of Southern Alligator Lizard's eyes are light yellow, compared with the darker eyes of the Northern alligator lizard. Typically, Southern Alligator Lizards have dark lines running lengthwise through the middle of the scales on their ventral side, while Northern Alligator Lizards will usually have dark lines, that also run lengthwise on the underside, but appear between the scales along their edges.

Each species may also occupy different habitats when found in the same location. Southern Alligator Lizards can often be found in drier and hotter habitats than Northern Alligator Lizards.

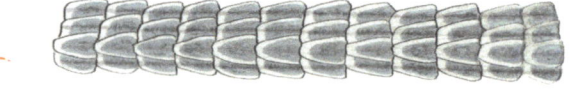

Southern (above) and Northern Alligator Lizard (below) ventral scales. Illustration by Sofia Vermeulen.

Southern Alligator Lizard
Elgaria multicarinata

Copulating Southern Alligator Lizards. Illustration by Sofia Vermeulen.

Description
This lizard ranges from 2.875 to 7 inches in snout to vent length. Dorsal coloration is brown, grey or yellowish above, with red blotches on the middle of the back. Some Southern Alligator Lizards have very pronounced dark dorsal bands, while others are covered with a reddish or yellowish coloration.

Behavior
This species prefers drier and hotter habitats than E. coerulea and is generally found near sunny clearings. Typically, Southern alligator lizards have dark lines running lengthwise through the middle of the scales on their ventral side. Hatchlings are small with smooth shiny skin and a plain tan, light brown or copper colored back and tail. The sides are darker and sometimes mottled or barred as they are on adults. Juveniles gradually develop the large scales and heavy dark barring found on the back and tails of adults. While mostly diurnal creatures, they can be crepuscular and nocturnal during hot weather and inactive during cold periods in winter. Eggs are laid from May to July and they hatch during late summer and early fall.

Southern Alligator Lizard in the process of regenerating a new tail.

Habitat
Southern Alligator Lizards usually frequent well-wooded areas and open annual grasslands on the reserve.

Interesting Facts
During the breeding season, a male may grab onto the head of a female with his mouth until she is ready to let him mate with her and they can remain attached this way for hours, seemingly oblivious to their surroundings. Besides keeping her from running off to mate with another male, this demonstrates how strong and suitable a mate the male lizard is.

Notes
I frequently had these lizards evade my grasp by dive bombing into the nearby grass, folding their limbs back and "swimming" through the grass to safety. While handling them, I found the lizards would twist their bodies in a corkscrew like motion to escape my grasp, smear their cloacal contents and (depending on the size) could deliver a painful bite. I often found them being parasitized by nymphal ticks around the neck and ear regions. I once found a pair of copulating Southern Alligator Lizards underneath a coverboard, neither gender seemed entirely thrilled at my interference and they both ran away together, while still in amplexus.

Northern Alligator Lizard
Elgaria coerulea

Northen Alligator Lizard found on UCSC campus at The Village.

Description

This lizard ranges from 2.75 to 5.875 inches in snout to vent length. Their head is usually not heavily mottled with dark color. Color is brown, grey, olive, or brown (above) with a broad band of olive-gray to brown down the middle of the back, sometimes with spots, and with darker sides mottled with dark spots. Usually there are dark lines running lengthwise on the underside, which run between the scales, along their edges, which (when present) is the most commonly used feature to distinguish this species from the southern species. Newborn lizards are very thin and small, roughly four inches long, with smooth shiny skin with a plain tan, light brown, or copper-colored back and tail. The sides are darker and sometimes mottled or barred as they are on adults. Juveniles gradually develop the large scales and heavy dark bars found on the back and tails of adults.

Behavior

Alligator Lizards move with a snake-like undulating motion, often tucking the rear legs up against the side of the body and pulling itself along on its belly with its front two feet. Alligator Lizards use their prehensile tail to climb trees in search of food and escape predation. The tail of an Alligator Lizard is easily broken off and will grow back, although generally not as well as the original. When first detached, the tail will writhe around for several minutes, long enough to distract a hungry predator from the lizard. Other defensive tactics used by Alligator Lizards are smearing the contents of the cloaca on the enemy and biting; males will also sometimes evert their hemipenes when threatened. They often bite onto a predatory snake, on the neck or the head, rendering the snake unable to attack and, when handled by people, will readily bite and can draw blood. One observer reported seeing a juvenile Alligator Lizard bite onto its own tail, making itself impossible to be swallowed by a juvenile Alameda Striped Racer, which eventually gave up! The Alligator Lizard is an example of a classic feeding generalist, consuming a wide variety of prey, including invertebrates, lizards, bird eggs and small mammals. During the spring breeding season, a male lizard grabs onto the head of a female with his mouth until she is ready to let him mate with her. They can remain attached this way for many hours, almost oblivious to their surroundings. Besides keeping her from running off to mate with another male, this probably shows her how strong and suitable a mate he is. This lizard is active during the day and inactive during colder periods in winter. Young are born live and fully-formed, sometime between June and September; this live birth allows for more parental control over the developing embryo's core temperature.

Habitat

This lizard prefers wetter and cooler habitats than Southern Alligator Lizards. They are likely found near Coast Live Oak Woodland and California Annual Grassland habitats.

Ventral shot of a gravid female Northern Alligator Lizard.

Interesting Facts
Mother Alligator Lizards have been observed tearing through the amnion to help free their young struggling to exit the membrane and consume the remnants of the amniotic sac to regain energy after the costly process of reproduction.

Notes
This was the only species that I didn't observe on the reserve during my surveys. Similar to Southern Alligator Lizards, I've found they'll often excrete their cloacal contents, aggressively bite and twist in a corkscrew motion to escape capture while being held. It's possible that this species is found on the reserve, but my surveying wasn't performed in the fall or winter, so this species was less likely to be observed in comparison with the Southern variety, which is more adept to the heat of springtime.

Northern California Legless Lizard
Aniella pulchra

California Legless Lizard. Illustration by Dillyn Adamo.

Description

This underground dweller ranges from 4.375 to 7 inches from snout to vent. A small slender lizard with no legs, distinct eyelids, a shovel-shaped snout, smooth shiny scales and a blunt tail. This lizard looks like a snake, but look for the eyelids (snake don't have them!) Their dorsal coloration varies from metallic silver, beige, dark brown, to black. Ventral coloration varies from whitish to bright yellow. Typically there is a dark line along the back and several thin stripes between scale rows along the sides where the dorsal and ventral colors meet, but variants occur.

Behavior

Legless lizards don't bask in direct sunlight. Their tolerance for low temperatures allows them to be active in cool conditions. This species lives mostly underground, burrowing in loose sandy soil, while foraging in loose soil, sand, and leaf litter during the day. Sometimes they're found above ground, at dusk and at night. Apparently, this lizard is active mostly during the morning and evening when they forage beneath the surface of loose soil or leaf litter which has been warmed by the sun.

Like other lizards, their tail can detach and writhes on the ground for several minutes to distract a potential predator while they escape. Known predators of the California legless lizard include Monterey Ring-necked Snakes, Common Kingsnakes, Deer Mice, domestic cats, California Thrashers, American Robins, and Loggerhead Shrikes, most of which have been seen on the reserve. Little is known about their reproduction other than the fact that they are live-bearing and probably breed between early spring and July, with one to four young (usually two) born between September and November.

This California Legless Lizard was found under the base of bush lupine on the FONR.

Habitat
Legless lizards occur in moist, warm, loose soil with plant cover, as well as in sparsely vegetated areas of California Annual Grassland, Maritime Chaparral and Oak Woodlands.

Interesting Facts
These lizards conceal themselves beneath leaf litter or substrate, then ambush their prey. They primarily eat larval insects, beetles, termites and spiders.

California Legless Lizard unearthed while digging through loamy soil underneath Bush Lupines at the Fort Ord Natural Reserve.

Notes
I unearthed this mysterious lizard a few times while sifting through soft, loamy, moist soil at the base of Bush Lupine, Sandmat Manzanita and California Sagebrush. Whenever I handled these lizards, they would writhe and bite in an attempt to break free and would often try to burrow into my hand to escape capture. In addition to the aforementioned behaviors, I found these lizards would often excrete their cloacal contents, presumably to deter predators. I attempted to attain photographic evidence of their eyelids (which is one of their distinguishing features from snakes), but they appear to close their eyelids only while underground.

Northern Pacific Rattlesnake
Crotalus oreganus oreganus

Northern Pacific Rattlesnake. Illustration by Willow Mosely.

Description

Adults typically range from 15 to 65 inches long from head to tail. This stocky pit viper has a large triangular head, elliptical pupils, keeled scales and a rattle on the end of the tail comprised of interlocking hollow segments. A new rattle segment is added to the anterior side each time the snake's skin is shed. The frequency that the snake sheds is influenced by diet and habitat preferences. Because of this variability between sheddings, trying to correlate a snake's age based off the number of rattles is inaccurate. The snake's dorsal coloration is variable, often matching the environment it inhabits and ranging from olive-green, gray, brown, golden, reddish-brown, yellowish or tan. In addition, this species has black blotched markings, usually with dark edges and light borders that mark the back, with corresponding blotches on the sides. The underside is pale to yellow and can appear weakly mottled. Young are born with a bright yellow tail with a single button (the beginning of a rattle), which does not make a sound.

Behavior

Rattlesnakes are primarily nocturnal and crepuscular during times of excessive heat, but also diurnal when the temperature is more moderate. Rattlesnakes are relatively inactive in colder areas, where they are known to den in burrows, caves, and rock crevices, occasionally found in large numbers or with other snake species.

Juvenile Rattlesnake with the original button still on its rattle.

When alarmed, a rattlesnake will quickly shake its tail back and forth using special tail musculature. The movement rubs each of the rattle segments together, producing a buzzing sound, which serves as a warning to nearby predators. A bite from this snake can be fatal without immediate medical treatment! Even a dead snake can "bite" and inject venom if the fangs are touched.

Loreal pits located on the sides of the head, which appear as a second set of nostrils, sense heat and help the snake locate prey by infrared detection. Rattlesnakes perceive their prey using their vision, their sense of smell, their ability to detect vibrations and their ability to sense heat. Rattlesnakes prey on a variety of organisms and will most often ambush and bite their prey from a location near a commonly used trail.

Once they inject their prey with venom, the snake releases, waits, and then follows the chemical trail of the envenomated animal (with help from the Jacobson's organ) and swallows it whole. Unlike other snakes which kill their prey via constriction, rattlesnakes use different concoctions made of either neurotoxic and/or hemotoxic venom to debilitate and consume their prey.

Rattlesnakes eat birds, lizards, snakes, frogs, insects, and small mammals. Some prey, such as adult California Ground Squirrels, have evolved immunity to rattlesnake venom and will vigorously bite and claw at any rattlesnake they feel to be a threat. Adult males are fairly territorial and can be observed engaging in a "combat dance" during the spring breeding season, where both male snakes intertwine and slam each other to the ground until the weaker snake concedes and leaves the area.

Rattlesnakes are ovoviparous, meaning that a pregnant female retains her fertilized eggs inside her body and gives birth to living young. Breeding typically occurs in the spring. Males must search extensively for females during the mating season, aided by the Jacobson's organ to help track any pheromones being released by receptive females. An average litter consists of 4 to 12 young, which are born from August to October.

Habitat

Northern Pacific Rattlesnakes inhabit the Coastal Scrub, Annual Grasslands, and a few intergrade regions with Coast Live Oak Woodlands.

Interesting Facts

Unlike certain endangered or threatened species, Western Rattlesnakes have no legal protection from people hunting them for sport or engaging in "rattlesnake roundups," which happen annually in other states and can harm local populations. Thankfully, this trend never gained popularity in California.

Notes

I experienced the classic defensive rattling with the only living individual I found during my surveys. I found it basking near a shady grove of oak trees frequented by ground squirrels. When disturbed, it coiled its body in a ball, with its tail raised, rapidly shaking back and forth, while slithering away to take refuge under a nearby California Sagebrush shrub. Although I've experienced the occasional rattlesnake confrontation before, most individuals I've observed in the past were merely basking or hiding among the brush and behaved in a way to remain unseen and sheltered from human interaction.

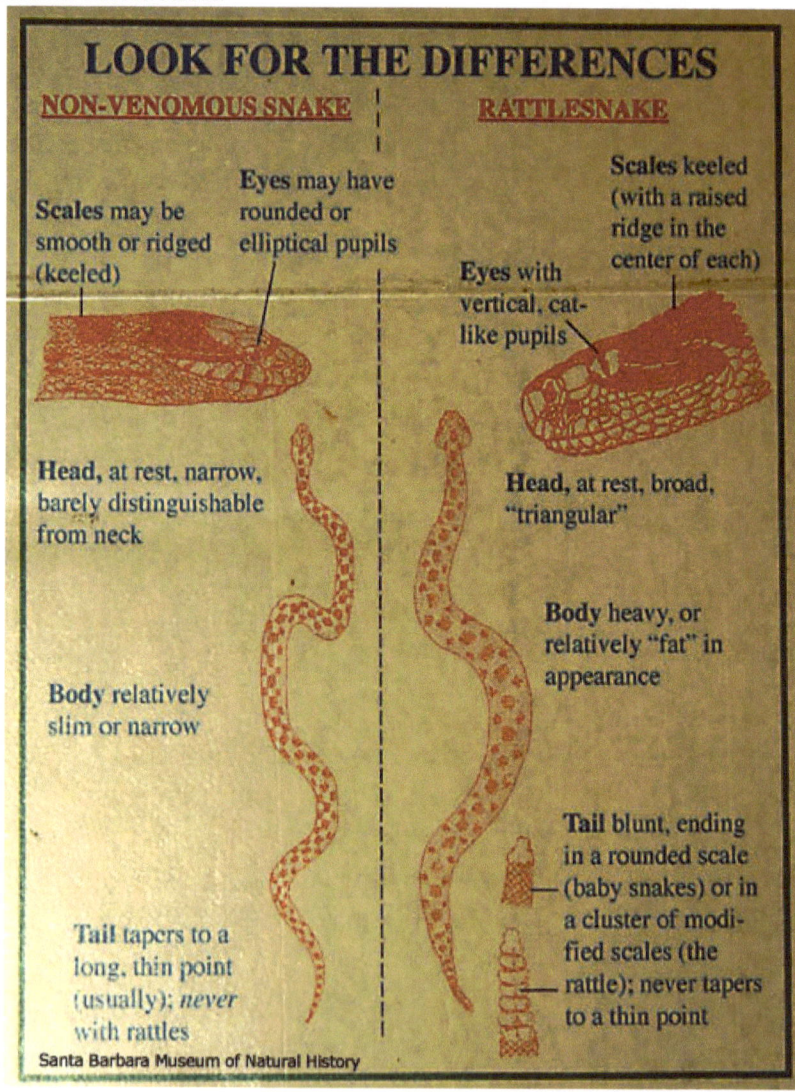

Photo taken from http://www.californiaherps.com/images/signs/rattlesnakedifferences.jpg

Coast Garter Snake
Thamnophis elegans terrestris

Coast Garter Snake mating ball. Illustration by Elexis Padron.

Description

Measuring 18 to 43 inches in length from head to tail, this medium-sized slender snake has a head barely wider than its neck with keeled dorsal scales. Their color and pattern is highly variable, but there is usually a yellow dorsal stripe and two yellowish lateral stripes along each side of its body. Their underside is yellowish to bluish-gray, with varying amounts of reddish markings. Certain Garter snakes in the Monterey Bay area have red side stripes, with varying degrees of checkering or barring of the black on a reddish ground color.

Behavior

Coast Garter Snakes are active in daylight and are chiefly terrestrial. This species is not as dependent on water as other Garter snake species, but still is more likely to be found near water. When frightened, they will sometimes seek refuge in vegetation or ground cover, but they'll also crawl quickly into water and swim away from trouble. If frightened when picked up, this snake will often strike repeatedly, release cloacal contents and musk. Coast Garter Snakes have toxins in their saliva, which can be deadly to their prey and even though their bite might produce an unpleasant reaction in humans, they aren't considered dangerous to people. Another classic example of a feeding generalist, this snake eats a wide range of prey (among the widest of any snake species), including amphibians and their larvae, fish, birds, mice, lizards, snakes, worms, leeches, slugs and snails. They breed primarily in spring, with young born live from July to September.

Coast Garter Snake found at Ano Nuevo State Park.

Habitat
Coast Garter Snakes inhabit the Coast Live Oak Woodlands and California Annual Grasslands on the reserve.

Interesting Facts
This snake has such intense pheromones, that a mating ball (writhing orgy of many snakes) will often form as a result of numerous sexually mature snakes clustering together.

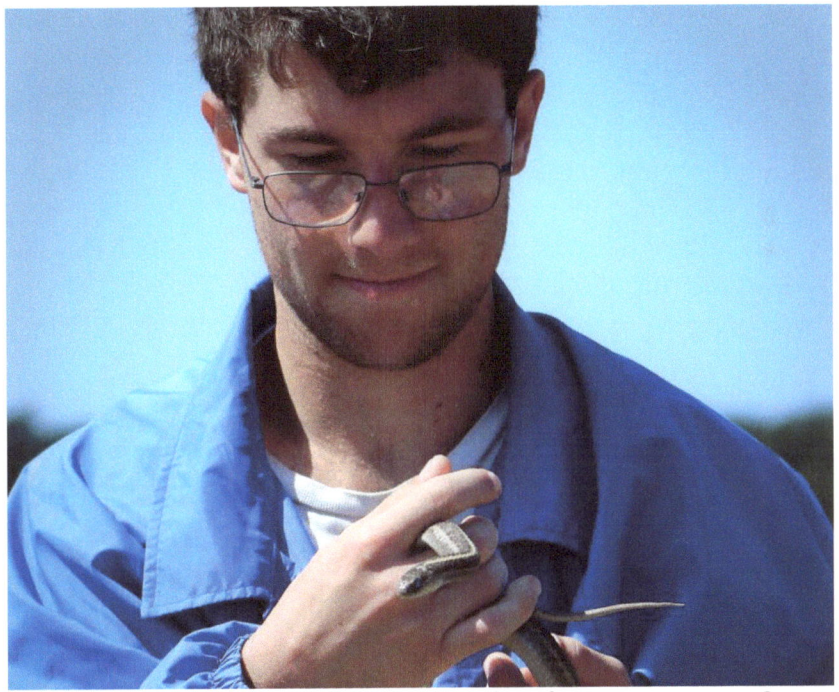

The author holding a Coast Garter Snake, caught while performing San Francisco Garter Snake surveys at Ano Nuevo State Park.

Notes
I didn't find any Coast Garter Snakes during my surveys, but its presence in the Coast Live Oak portions of the reserve has been confirmed by other authorities. In other places up the coast, I've frequently observed this species elsewhere inhabiting the Mixed Oak Woodlands and Annual Grasslands, often clustered together underneath the same object. I've found these snakes will almost always musk on contact and will occasionally bite if they feel threatened.

Adult male Western Yellow-bellied Racer, found copulating underneath a coverboard in the Coastal Scrub in late April.

Western Yellow-bellied Racer
Coluber constrictor mormon

Juvenile (top) and adult (bottom) Racer. Illustration by Hannah Caisse.

Description

Adult Racers vary between 20 to 75 inches long from head to tail, typically less than 3 feet long, while hatchlings are from 8 to 11 inches long. This slender snake has large eyes, broad head, slender neck, smooth scales, and a long thin tail. Overall their dorsal coloration varies from brown, blue-grey, or olive to green. This contrasts with a white or yellowish venter. Juveniles look significantly different than adults, with dark blotches on the sides and saddled markings on the back. They are often mistaken for young Gopher Snakes.

Behavior

Racers are diurnal and can be seen hunting with their head held high off the ground, sometimes moving it from side to side. They are often found at denning sites along with other species of snakes. Mostly terrestrial, the Western Yellow-bellied Racer is also a good climber and has been found in trees occasionally. Despite the species' latin name (*Coluber constrictor*), they are not constrictors. Their prey are killed by being quickly overcome, captured, and crushed with their jaws or trapped under their body and swallowed alive. This species is a feeding generalist and eats a variety of organisms such as lizards, small mammals, birds, eggs, snakes, small turtles and frogs, as well as large insects. Cannibalism has also been observed. They lay 3 to 11 eggs in mid-summer, sometimes in communal nests and their eggs hatch in late summer.

Juvenile Racer. Illustration by Hannah Caisse.

Habitat

Racers prefer open areas with sunny exposure on the reserve. This includes Annual Grassland, Coastal Scrub, Maritime Chaparral, and Oak Woodlands.

Interesting Facts

At one time, juveniles were thought to be a different species of snake than the adults due to the dramatic difference in their appearance.

Juvenile Racer found scurrying through the Annual Grasslands of FONR. Compare with Gopher Snake photograph.

Notes

I found this snake can be somewhat aggressive and has a tendency to bite when handled and will essentially always musk on first contact. I've definitely mistaken juvenile Racers for young Gopher Snakes at first glance, but usually tell them apart based off the former's smooth dorsal scales, wedged preoculars, large eyes, and dark iris. I once found a pair of racers copulating under a coverboard in the Coastal Scrub. When approached, the female bolted while the male remained and coiled into a defensive pose.

Ring-necked Snake found seeking refuge under downed wood in the Oak Woodland.

Monterey Ring-necked Snakes coil their tails and display the brightly colored ventral side to direct a predator's strike away from the head and vital organs.

Monterey Ring-necked Snake
Diadophis punctatus vandenburghii

Monterey Ring-necked Snake. Illustration by Stacy Wu.

Description

The typical total length of an adult Monterey Ring-necked Snake can vary widely, but is generally about 11 to 16 inches, while the record length is 33-5/8 inches. A small, thin snake with smooth scales and gray, blue-gray, blackish or dark olive dorsal coloring, with a bright orange to reddish underside, speckled with a few small black markings. A narrow orange band circles the neck, which readily distinguishes this species from other snakes.

Behavior

Although this snake is mildly venomous, it's not considered dangerous to humans. Ring-necked Snakes have enlarged teeth in the rear of the upper jaw and mild venom, which helps to incapacitate small prey. They typically feed on small salamanders, tadpoles, small frogs, small snakes, lizards, worms, slugs and insects. This species lays its eggs in the summer, sometimes in a communal nest.

Habitat

The Monterey Ring-necked Snake species is rather secretive and is typically found in the shadier groves of the Coast Live Oak Woodland, California Annual Grassland, and Maritime Chaparral habitats of FONR.

A Large Ring-necked Snake observed basking in the middle of the road.

Interesting Facts

When disturbed, it coils its tail like a corkscrew, exposing the bright red underside as a warning display and may also smear musk and other cloacal contents to discourage its consumption.

Notes

Nearly every single time I encountered Monterey Ring-necked Snakes, they would coil their tails in a corkscrew, expose their brightly colored ventral side and excrete and smear (and occasionally fling) their particularly foul-smelling musk on me. I even had one individual fling musk everywhere, some of which hit me in the face and was incredibly repulsive. After experiencing this defense mechanism, I can understand how effective it would be at deterring predation!

Common Sharp-tailed Snake
Contia tenuis

Common Sharp-tailed snake. Illustration by Mariam Moazed.

Description

Adult Sharp-tails average 8 to 12 inches from head to tail, with some nearly 18 inches long. This snake has a small thin body and a sharp point on the end of the tail. The head of an adult is typically medium to light olive-gray or brown with black flecking or blotches, occasionally with orange blotches. Their dorsal coloration is rusty-red or orange-red, while irregular black bands mark the ventral side. Each ventral scute is marked with one band, with the bands becoming faint or absent towards the tail, and absent from the anal plate and the caudal scales. Juveniles typically have brighter dorsal coloration than adults.

Sharp-tailed Snake black banding on ventral side. Illustration by Mariam Moazed.

First confirmed sighting of a Sharp-tailed Snake on the FONR: found below a coverboard, beneath a Coast Live Oak in late winter.

Behavior

This snake is an excellent burrower and spends a significant amount of time underground. Sharp-tailed Snakes require moist soil and are often encountered underneath surface objects in open, grassy areas near forests. They are active when the ground is damp, occasionally during or after rains and sometimes when surface temperatures are as low as 50 degrees. This snake feeds on slugs and their eggs and on slender salamanders and has relatively long teeth, which allow the snake to hold onto slippery prey. Sharp-tailed Snakes will lay eggs in June or July.

Habitat

This snake is relatively secretive and is thought to be found in the oak woodland, annual grassland, and maritime chaparral parts of the reserve.

The author holding his first finds from the FONR coverboard surveys: on the left is a Sharp-tailed Snake and on the right is a Southern Alligator Lizard.

Interesting Facts

Sharp-tailed Snakes use their pointed tails to pin down and stabilize small prey, such as slugs for consumption.

Notes

I observed this species a few days after a long rain on a nice sunny day in late winter. I found it alongside a young Southern Alligator Lizard and some slugs, which I suspect it was going to feed on. I saw very few Slender Salamanders on the reserve which might account for the lack of Sharp-tailed Snakes seen in the area as well.

Pacific Gopher Snake
Pituophis catenifer catenifer

Gopher Snake. Illustration by Hannah Caisse.

Description
Adult Gopher Snakes can be 2.5 to 9 feet long, although most adults of this subspecies usually range from 4.5 to 5 feet long. This large snake has heavily keeled scales, a narrow head, slightly wider than its neck and a protruding rostral scale on the tip of its rounded blunt snout. Ground color is straw or tan, with large square dark chocolate blotches or saddles along the back and smaller gray spots on the sides. The back of the neck is dark brown and the venter is cream to yellowish with dark spots. Often there is a reddish color on the top, especially near the tail. There is usually a dark stripe across the head in front of the eyes and a dark stripe from behind each eye to the angle of the jaw.

Behavior
Gopher Snakes are often mistaken for rattlesnakes and killed unnecessarily. They are also frequently run over by vehicles when crossing roads. Gopher snakes are mostly diurnal and seen burrowing, climbing or even swimming. They are one of the most commonly observed snakes on roads and trails, especially in the spring when males are actively seeking a mate and in the fall when hatchlings emerge.

When threatened, a Gopher Snake will elevate and inflate its body, flatten its head into a triangular shape, hiss loudly, and quickly shake its tail back and forth to make a buzzing sound. This may be a mimic of an alarmed rattlesnake or it could be a similar behavior that helps to warn off an animal that could be a threat. Gopher Snakes eat mostly small mammals, especially pocket gophers, moles, rabbits, and mice, along with birds and their eggs and nestlings. They occasionally consume lizards and insects. A powerful constrictor, these snake kill their prey by suffocating them in their body coils or by pressing the animal against the walls of their underground burrows. Females lay one to two clutches of two to 24 eggs from June to August.

Habitat

This species is found in a variety of habitats on the reserve:, including Annual Grassland, Oak Woodland, Mixed Chaparral, and Coastal Scrub.

Young Gopher Snake found basking in the sun at Fort Ord. Compare with photo of Juvenile Racer.

Gopher Snake blending in with its surroundings.

Interesting Facts

Gopher Snakes have a specially developed epiglottis, which augments the sound of their hiss when air is forced through the glottis.

Notes

I've been startled on occasion by these snakes while hiking around the reserve, initially mistaking them for rattlesnakes after they strike a pose. Several times I have caught these snakes by the tips of their tails, just as they were diving into a gopher hole. One time, I unearthed a gopher snake from a hole that was in the process of consuming a rodent, with just the tail of the rodent sticking out of its mouth.

California Kingsnake
Lampropeltis getula

California Kingsnake. Illustration by Willow Mosely.

Description
California Kingsnakes seldom exceed 48 inches and are most commonly 2.5 to 3.5 feet in length, with hatchlings around 12 inches long. This snake has smooth, shiny, unkeeled scales with a head that's barely wider than the neck. They are most commonly seen with alternating bands of black or brown and white or light yellow, including the underside, where the light bands become wider.

Behavior
Kingsnakes are active during daylight in cooler weather and at night when temperatures are higher. When disturbed, Kingsnakes are generally not aggressive, but sometimes they vibrate their tail quickly, hiss and roll into a ball to hide their head and show their vent. They eat a wide variety of prey, including rodents and other small mammals, lizards, lizard eggs, snakes (including Rattlesnakes), snake eggs, turtle eggs and hatchlings, frogs, salamanders, birds, bird eggs and chicks and large invertebrates. Kingsnakes are powerful constrictors and immune to rattlesnake venom and have been known to consume this otherwise deadly adversary. California Kingsnakes are oviparous, laying eggs that incubate and then hatch.

Courtship behavior between a male and a female involves the male neck-biting to hold the female during copulation. One to two months after breeding, females lay anywhere from 3 to 24 eggs, averaging a clutch size from 8 to 10, with egg-laying occurring generally between May and August.

Kingsnake found seeking shelter underneath a coverboard.

Habitat

I observed this species on the southern plot of the reserve in sparsely vegetated Annual Grasslands, where they were presumably feeding on the abundance of Fence lizards, birds and rodents in the area.

Interesting Facts

Kingsnakes are immune to rattlesnake venom and can occasionally be observed consuming these otherwise deadly foes.

Notes

When I handled this snake, it often musked on first contact and occasionally vibrated its tail back and forth similar to a rattlesnake's defense. This snake replaced a pair of Western Skinks, that were living underneath a board consistently for four weeks; I suspect the Kingsnake ate the previous inhabitants and settled into its new home. Once while out herping off the reserve, I was initially startled by the rattle of a rattlesnake only to investigate further to see a kingsnake in the process of consuming it!

Western Pond Turtle
Actinemys marmorata

Western Pond Turtle. Illustration by Sasha Taus.

Description
Typical adults are 3.5 to 8.5 inches in shell length, with hatchlings approximately one inch in shell length, with their tail nearly as long as their shell. These turtles are drab, dark brown, olive brown, or blackish, with a low unkeeled carapace, usually with a pattern of lines or spots radiating from the centers of the scutes. The legs and head have black speckling against a cream to yellowish background color. Males usually have a light throat with no markings, a flatter shell and a concave plastron (shell underside), while females usually have a throat with dark markings, a taller shell, and a flat or convex plastron, which tends to be more heavily patterned than males.

Behavior
Western Pond Turtles are both diurnal and aquatic and often seen basking above the water, but will quickly slide into the water when they feel threatened. Pond turtles hibernate underwater for several months during the winter and will cluster in the shallow ends of the ponds. They eat aquatic plants, invertebrates, worms, frog and salamander eggs and larvae, crayfish, carrion and the occasional frog and fish. Between April and August, reproductive females will migrate along stream or pond margins, where they lay a clutch of 2 to 11 eggs.

Western Pond Turtle found in Los Padres National Forest.

Habitat

As their name suggests, the Western Pond Turtle is typically found near bodies of water with abundant vegetation and either rocky or muddy bottoms, in woodland, forest and grassland. They prefer pools to shallower areas and require logs, rocks or exposed banks for basking.

Interesting Facts

Reproductive females will migrate up to a kilometer or more onto land to dig a nest and lay their eggs.

Notes

This species has been found on the reserve in previous years, but was not observed during my eleven weeks of surveying. I suspect this species migrates onto the reserve from vernal pools in the nearby agricultural areas or Bureau of Land Management (BLM) regions, which often contain the necessary ponds this species requires for survival.

Pacific Chorus Frog
Pseudacris regilla

Pacific Chorus Frog. Illustration by Tina Milz.

Description
  Adults are 0.75 to 2 inches long from snout to vent. This is a small frog with a large head and eyes, slim waist, round pads on the toe tips, limited webbing between the toes, and a wide dark stripe through the middle of each eye that extends from the nostrils to the shoulders. Often there is a Y-shaped marking between the eyes. Dorsal coloration is highly variable, including green, tan, brown, gray, reddish or cream colored. Most often they are green or brown. Their underside is creamy with yellow underneath their back legs. In response to changing environmental conditions, Pacific Chorus Frogs can quickly change between dark and lighter coloring. Sometimes dark markings on their back and legs can vary in intensity or disappear during this transformation.

Behavior
  The Pacific Chorus Frog is active both day and night, becoming mostly nocturnal during drier periods. Their old name, Pacific Tree Frog, highlighted this species' habit of occaionally climbing high in vegetation. However, this frog is primarily a ground-dweller, living among shrubs and grass typically near surface water.

Chorus Frogs' large toe pads allow them to climb easily and cling to branches, twigs and grass. Green body color absorbs more solar radiation, which can be more beneficial in cold and aquatic habitats. Brown body color absorbs less solar radiation, which may be more beneficial in drier, hotter, more terrestrial habitats. Chorus Frogs reproduce in water. Fertilization is performed externally, with the male grasping the back of the female and releasing sperm into the surrounding water as the female lays her eggs. Adult males call to advertise their fitness to females to pair up in amplexus in the water. Once eggs are laid, the adults leave the water and the eggs hatch into tadpoles, which feed in the water. Each tadpole will eventually grow four legs, lose its tail and emerge onto land where they disperse into the surrounding territory. Breeding and egg-laying occurs between November until July, depending on the location. Females lay on average between 400 to 750 eggs in small, loose, irregular clusters of 10 to 80 eggs each. Egg clusters are attached to sticks, stems or grass in quiet, shallow water. Tadpoles will congregate for thermoregulatory and anti-predation purposes. They metamorphose in about 2 to 2.5 months, generally from June to late August.

Pacific Chorus Frog camouflaging in with its surroundings.

Pacific Chorus Frog found in the author's backyard.

Habitat

This species utilizes a wide variety of habitats, often far from water outside of the breeding season, including Oak Woodland, Mixed Chaparral, and Annual Grassland.

Interesting Facts

Chorus Frog calls are probably the most recognizable frog call worldwide as a result of their prolific use as background sounds from countless night scenes in hollwood movies. As a result, this frog has been dubbed the "Hollywood Frog."

Image from Chuck Jones' "One Froggy Evening" taken from https://www.cartoonbrew.com/wp-content/uploads/2005/10/onefroggyevening_post.jpg

Notes

I only had one Pacific Chorus Frog observation during my 11 weeks of surveying. Even though I didn't observe any persistent bodies of water on the reserve, there must have been at least one basin on, or adjacent to, the reserve that lasted long enough to raise tadpoles, which proliferated throughout the reserve. As for now, the origin of the Chorus Frogs on the reserve will remain a mystery. I suspect I didn't find many Pacific Chorus Frogs during my surveying in part because of the general absence of water on the reserve, but also because I didn't survey past sunset, which is when Chorus Frogs are likely most active.

Plethodontidae

This is the largest family of salamanders, containing close to two thirds of all known species. Plethodontid salamanders (Climbing Salamanders, Slender Salamanders, and Ensatinas) don't have lungs and breathe through their skin, which requires them to live in moist environments on land (typically underground) and to move about only during periods of high humidity. Plethodontid salamanders are also distinguished by their nasolabial grooves, which are vertical slits on either side of the face, between the nostrils and upper lip that are lined with chemosensory glands.

All Plethodontid salamanders native to California lay eggs in moist places on land, where their young develop and hatch directly into tiny terrestrial salamanders with the same body form as an adult. Unlike most amphibians, they do not hatch in the water or begin their lives as tiny swimming larvae breathing through gills.

Monterey Ensatina found seeking shelter underneath a coverboard.

Monterey Ensatina
Ensatina eschscholtzii eschscholtzii

Monterey Ensatina. Illustration by Dillyn Adamo.

Description
Adult Ensatinas measure from 1.5 to 3.2 inches long from snout to vent. This medium-sized salamander has long legs and a relatively short body, with 12 to 13 costal grooves. A subtle constriction at the base of the tail differentiates ensatinas from other similar species. Like all plethodontids, Ensatinas have nasolabial grooves and no lungs. The eschscholtzii subspecies is reddish brown to pinkish brown above, and whitish below, with orange or reddish-orange skin at the base of all four limbs. Their eyes are very dark with no yellow markings. Males have longer, more slender tails than females and a shorter snout with an enlarged upper lip. The bodies of females are usually shorter and fatter than the males.

Behavior
Ensatinas live in relatively cool, moist places on land, becoming most active on rainy or wet nights when temperatures are moderate. They stay underground during hot and dry periods where they are able to tolerate considerable dehydration and can continue to feed during the summer months. When severely threatened, an Ensatina may drop its tail to distract the attention of a predator towards the writhing tail, so the animal can crawl away to safety. Ensatinas eat a wide variety of invertebrates, including worms, ants, beetles, spiders, scorpions, centipedes, millipedes, pill bugs and snails.

Ensatinas expel a surprisingly long, sticky tongue from their mouth to capture prey and pull it back into their gullet where it is crushed and swallowed. Typically, feeding is accomplished using sit-and-wait ambush tactics, but ensatinas will sometimes slowly stalk their prey. Rarely, Ensatinas can make a hissing sound, similar to the hissing of a snake, when threatened.

Unlike some salamanders, Ensatinas mate on land rather than in water. Breeding takes place in fall and spring, but may also occur throughout the winter. Ensatina courtship involves an elaborate ritual, where the male rubs his body and head against the female, eventually dropping a sperm capsule onto the ground, which the female then picks up with her cloaca. Females lay 3 to 25 eggs, with 9 to 16 being average. Unlike most other reptiles and amphibians, female Ensatinas typically remain with their eggs and will guard them until they hatch.

Monterey Ensatina discovered in the Maritime Chaparral.

Habitat

Monterey Ensatinas inhabit the Coast Live Oak Woodlands, Annual Grasslands, and Maritime Chaparral on the reserve and can be found in gopher holes and under rocks, logs, and other debris, especially bark that's fallen off decaying logs and trees.

Monterey Ensatina found inside a piece of rotting driftwood.

Interesting Facts

When disturbed, an Ensatina will stand tall in a defensive posture with the tail extended up in the air and secrete a milky white substance from their tail, swaying it from side to side. The milky substance, acts like glue to temporarily seal the mouth of a potential predator while the salamander escapes. This toxic substance repels most predators, but some experienced predators have learned to eat all but the tail of Ensatinas! If a person gets this poisonous substance on their lips, they can experience some numbness for several hours.

Monterey Ensatina secreting milky substance from the tail. Photo taken by Gary Nafis from http://www.californiaherps.com/salamanders/pages/e.e.eschscholtzii.html

Notes

Before handling this organism, I always make sure to coat my hands in a layer of dirt. Because ensatinas breathe through their skin, it is important to prevent any oils from human skin from clogging their skin pores.

Ensatinas were by far the most common amphibian I found on the reserve, although it's likely that Slender Salamanders are the most common amphibian on the reserve, but weren't observed as often due to their fossorial behavior and camouflage. I occasionally found Ensatinas in remarkably dry portions of the reserve. Mainly, I found them in moister settings, underneath rotting logs and underneath oak duff. I even observed an Ensatina squeaking, even though they don't have lungs. I assume the mechanism for making this sound may be similar to how Arboreal Salamanders create their squeaking noises.

Gabilan Slender Salamander
Batrachoseps gavilanensis

Gabilan Slender Salamander. Illustration by Eleni Christoforou.

Description
Adults are 1.5 to 2.6 inches long from snout to vent. This small salamander with very short limbs, a long slender body with 19 to 22 costal grooves, and a narrow head and a long tail, give this species the worm-like appearance typical of most Slender Salamanders. If you look carefully, slender salamanders have only four toes on their front and hind feet. All other California Salamanders have five toes on their hind feet. Dorsal color is generally dark brown or gray above, with a red, brown, yellow, or tan dorsal stripe. Their venter (belly) is dark, with fine white speckles.

Behavior
These ubiquitous salamanders are chiefly active on rainy or wet nights when temperatures are moderate. During the long summer dry period, slender salamanders retreat underground. Individuals typically stay within a small range most of their lives, rarely moving more than a few feet. Slender Salamanders use several defense tactics, including coiling and remaining still, relying on their camouflaged bodies to avoid detection, uncoiling quickly, sprinting away and then remaining still again to avoid detection as well as dropping their tails, which wriggle on the ground to distract a predator long enough for it to escape.

Their diet consists of a variety of invertebrates, including small beetles, snails, mites, and spiders. Primarily a sit-and-wait predator, they catch their prey using a projectile tongue. Female slender salamanders lay eggs in October and November, shortly after the beginning of the fall rains. Clutch sizes range from 4 to 13 eggs on average. Their eggs are typically deposited in moist areas under objects such as rocks and logs or underground.

Slender Salamander remaining motionless to avoid detection.

Habitat

These salamanders were mostly found in the Coast Live Oak Woodland and different intergrade regions with California Annual Grassland and Maritime Chaparral on the reserve.

Slender Salamanders can often be found nesting together underneath a single stone or piece of driftwood.

Interesting Facts

Slender Salamanders have the capacity to curl into a ball and then spring into action, moving astonishingly fast for their small size.

Many Slender Salamanders housed together in the cavity of a rotting oak log.

Notes

Before handling this organism, I always make sure to coat my hands in a layer of dirt. Because Slender Salamanders breathe through their skin, it is important to prevent any oils from human skin from clogging their skin pores.

Surprisingly, I had very few encounters with this species on the reserve, even though it's probable that Slender Salamanders are the most common amphibian at FONR. This is likely because they possess excellent camoflage and spend most of their time underground. Their low sightings may likely be related to the lack of Sharp-tailed Snake sightings (a common predator).

I found slender salamanders buried in oak duff and underneath decaying logs, often well-camouflaged and clustered together within a single tree cavity or underneath a single log. I've found this species is much more active during the winter, but will remain underground during hot, dry periods.

Slender Salamander curling into a defensive posture.

Arboreal Salamander
Aneides lugubris

Arboreal Salamander displaying it's prehensile tail. Illustration by Elexis Padron.

Description

Adults measure 2.25 to 4 inches long from snout to vent and up to seven inches in total length. The Arboreal Salamander is a medium-sized lungless salamander with squarish toe tips, nasolabial grooves, around 15 costal grooves, and a prehensile tail. Dorsal coloration is reddish-brown, with small yellow spots, while the venter is creamy white, with light yellow undersides of the tail and feet. Young are dark, with gray or brass-colored patches. Males have broader, more triangular heads than females. This species is nocturnal and active when soil moisture is high after the onset of fall rains, but relatively tolerant of drier conditions compared to many other California salamander species. Its strong jaws and sharp teeth are capable of producing a painful bite.

Arboreal Salamander foraging in Coast Live Oak duff.

Behavior

Arboreal Salamanders are adapted for climbing, with long toes and a round tail in cross-section, which provide better grip and more flexibility to wrap their tail around branches. They have been found nearly 60 feet high in trees! Anti-predation behaviors include biting, a raised defensive posture, fleeing rapidly, and making a squeaking sound. They eat a variety of small invertebrates including millipedes, worms, snails, ants, termites, pill bugs, moths and centipedes, as well as Slender Salamanders. Their prey are captured by their projectile tongue and brought into the mouth, where they are crushed. A sit-and-wait-predator, adults forage for small invertebrates and sometimes Slender Salamanders on the ground at night during wet weather. Arboreal Salamanders perform a similar complex mating ritual akin to that of ensatinas, involving spermatophore exchange.

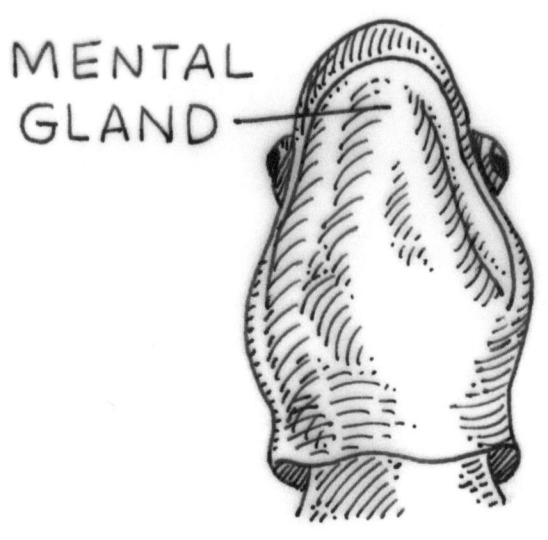

The Mental Gland is present in sexually mature plethodontid males. Mental Gland Illustration by Sasha Taus.

Breeding males have a heart-shaped mental gland under their chin and will put this gland on a female's back, stroking her back quickly with it during courtship. While scratching her skin with his teeth, the male delivers the mental gland pheromones (an aphrodisiac) to the female. In late spring and early summer, females lay from 5 to 24 eggs in moist places, most commonly in decaying cavities of Live Oak trees. Sometimes eggs can be laid high off the ground or under rocks and inside logs. Females usually remain with the eggs until they hatch, often coiled around them. On occasion, the eggs of several salamanders are found together in large masses.

Habitat

Arboreal Salamanders live in moist places on land, mostly in the Coast Live Oak Woodlands, but also in other drier habitats, including the Coastal Scrub and Maritime Chaparral of FONR.

Interesting Facts

Even though Arboreal Salamanders are lungless and breathe through their skin, they are capable of producing a squeaking sound by retracting their eyes into their sockets repeatedly when caught. When the eyeballs are depressed, their undersides protrude into the mouth cavity, thereby compressing the air in the mouth and forcing it outward in a squeaking sound!

Adult Arboreal Salamander hiding under a log.

Notes

Before handling this organism, I always make sure to coat my hands in a layer of dirt. Because these salamanders breathe through their skin, it is important to prevent any oils from human skin from clogging their skin pores.

I've found that most individuals of this species have a common tendency to bite when disturbed. They seem to prefer moister, colder conditions and I've often found them under decaying oak logs and duff and occasionally observed them in Mixed Chaparral habitat seeking shelter underground in gopher holes or dwelling in tree caverns. Although I've observed them active during the day, most are likely nocturnal.

California Tiger Salamander
Ambystoma californica

California Tiger Salamander. Illustration by Tina Milz.

Description
Adults are 3 to 5 inches long from snout-to-vent. A medium-large salamander with 12 prominent costal grooves, a short, rounded head, blunt snout, small eyes, nasolabial grooves absent and a tail flattened from side to side to help it swim. The California Tiger Salamander (unlike the other three salamanders found on the reserve) has an aquatic juvenile phase and a terrestrial adult stage where they breathe with lungs. Adults are a lustrous black with large yellow spots, often not present along the middle of the back. Larvae are yellowish-gray with broad caudal fins that extend well onto the back, broad flat heads and bushy external gills.

Behavior
California Tiger Salamanders are both nocturnal and spend most of their time underground in animal burrows, especially those of California Ground Squirrels. An active population of burrowing mammals is necessary to sustain sufficient underground refuge for these salamanders since burrows that are not maintained will eventually collapse.

This salamander needs both suitable terrestrial habitat with mammal burrows for refuge and temporary breeding ponds in order to survive. Adults probably feed mainly on a variety of invertebrates, while hatchlings feed on zooplankton and older larvae feed on tadpoles and other aquatic invertebrates. Predators include California Red-legged Frogs, American Bullfrogs, Garter Snakes, Skunks, and Ground Squirrels. Adults engage in a mass migration to breeding ponds during a few rainy nights, usually between November to May. During years without sufficient rainfall, migrations and breeding will not occur. Females lay eggs and attach them to underwater vegetation. One study showed females contained anywhere from 413 to 1,340 eggs, averaging 814.

Monterey County adult hybrid cross between the Barred Tiger Salamander and the California Tiger Salamander. Photo taken by Gary Nafis from http://www.californiaherps.com/salamanders/images/acalifatmavcrossmont506.jpg

Habitat

The California Tiger Salamander typically frequents Annual Grasslands and Coast Live Oak Woodland habitats.

Interesting Facts

Abystoma larvae, a.k.a. "waterdogs" are frequently used by anglers as bait for bass fishing. As a result of their high demand among bass fishers, local ponds throughout California have been perpetually stocked with non-native Tiger Salamanders since the 1950's.

California Tiger Salamander. Photo taken by Gary Nafis from http://www.californiaherps.com/salamanders/images/acalifornensesc.jpg

Notes

This species is protected by California state and federal laws and is considered threatened with possible extinction. Many populations have been extirpated due to habitat loss or fragmentation of suitable habitat through urbanization and agriculture. Eradication of California Ground Squirrels due to concerns about their effect on cattle grazing and agriculture may also threaten populations of this salamander because of its reliance on ground squirrel burrows. Predation by non-native bullfrogs also appears to be a serious threat. Perhaps the most tangible threat to California Tiger Salamanders is the proliferation of non-native Tiger Salamanders (*Ambystoma mavortium*) by bait dealers. Populations of these invasive Tiger Salamanders are well-established in many places throughout central California, and are eliminating *A. californica* through hybridization, competition for similar resources, and predation as larvae. This species has been found on the reserve in previous years, but was not observed during my 11 weeks of surveying.

I suspect this species migrates onto the reserve from nearby agricultural areas or Bureau of Land Management (BLM) lands, which often contain the necessary ponds this species requires for reproduction.

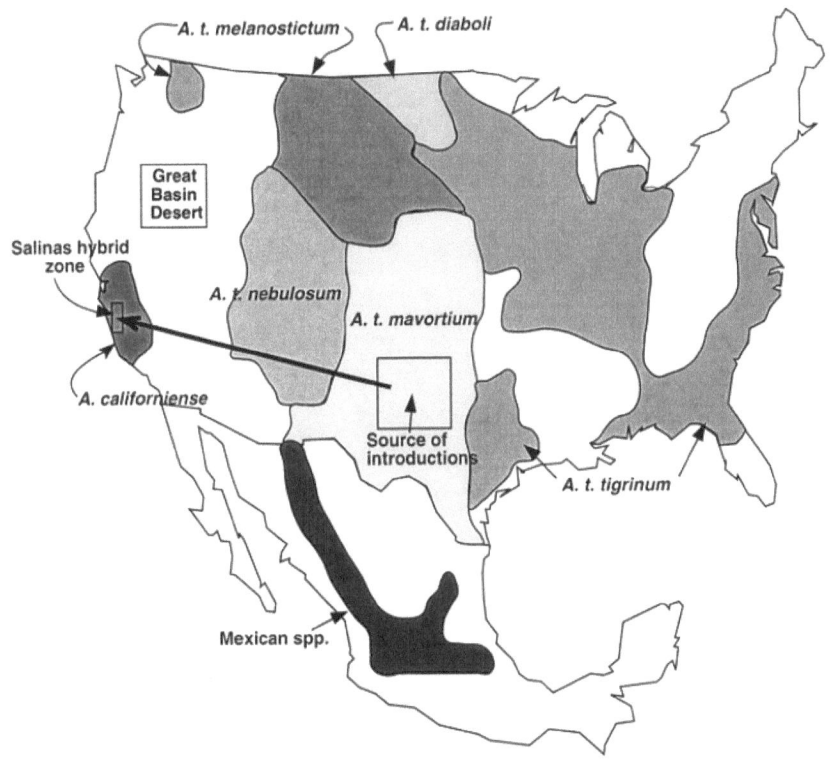

This figure shows the origins of the introduced Tiger Salamanders threatening California Tiger Salamanders. Figure taken from http://onlinelibrary.wiley.com.oca.ucsc.edu/doi/10.1890/02-5023/full

Acknowledgements

First and foremost, thanks to my wonderful family for fostering my passion and love of natural history, whether it was going on annual camping trips, frequently visiting zoos and museums, giving me taxidermied specimens as birthday presents or putting up with me housing fire-bellied frogs, red-eared sliders, bearded dragons and the occasional Slender Salamander or Kingsnake. All their support encouraged me to learn more about natural history. I'd also like to acknowledge the late Steve Irwin for inspiring a love of reptile and amphibians in me as a child. Thank you Gary Nafis for providing an invaluable resource (CalHerps) and allowing the use of your spectacular photos! Thank you Sasha Taus, Stacy Wu, Charlotte Grenier, Tina Milz, Hannah Caisse, Mariam Moazed, Sofia Vermeulen, Elexis Padron, Willow Mosely, Dillyn Adamo, Grace Ackles, Eleni Christoforou, and Juniper Harrower for providing me with fantastic illustrations for use in my guide. Thanks to Robert C. Stebbins, Samuel M. McGinnis, Alex Jones, Anne Bikle and Winifred Fick for making such amazing field guides, which acted as the templates for my field guide and the primary references for the majority of the information found within my guide. I'd like to thank Caleb Perez and Brandon Cluff for their help conducting the coverboard surveys. I also wanted to give special thanks to Chris Lay, Gage Dayton, Joe Miller, Patrick Robinson, Rob Burton, Emily Taylor, and Tony Frasier for helping inspire and facilitate my career in field work and natural history.

Coast Live Oak Woodland and California Annual Grassland at FONR.

Glossary

Amnion: Thin, transparent membrane surrounding newly-born animals.
Amplexus: The sexual embrace of a male amphibian around a female's chest or waist area.
Carapace: The hard, upper shell of a turtle, crustacean or arachnid.
Caudal Autotomy: The ability to intentionally remove one's tail via special musculature typically used as a mechanism among certain lizards and salamanders to escape predation.
Cismontane: On this side of the mountains.
Cloaca: The cavity at the end of the digestive tract for release of excretory and genital products in vertebrates (excluding most mammals) and some invertebrates.
Costal Folds: Vertical skin folds on the sides of some salamander bodies, separated by costal grooves.
Costal Grooves: Vertical furrows on the sides of some salamander bodies, separated by costal folds.
Coverboard: An artificial habitat, which creates an optimal microclimate meant to attract target species, in order to make it easier to capture them and collect data about the species of interest. At FONR, the coverboards used in the surveys were 4' x 4' x 1/2" plywood.
Crepuscular: An animal that is most active at dusk, dawn, or both.
Cycloid Scales: Scales with smoothly-rounded, free posterior borders.
Dichotomous Key: A key used to identify a plant or animal in which each stage presents descriptions of two distinguishing characters, with a direction to another stage in the key, until the species is identified.
Dorsal side: Of, on, or relating to the upper side or back of an animal, plant, or organ.
Epiglottis: A thin plate of flexible cartilage in front of the glottis that folds back over and protects the glottis during swallowing.
External Fertilization: When a male and female's gametes unite outside the female's body.

Femoral Pores: Pores on the underside of the thighs of certain lizard species that secrete a waxlike substance.
Forb: An herbaceous flowering plant that is not graminoid, especially one growing in a field, prairie or meadow.
Fossorial: A burrowing or underground life.
Glottis: The opening between the vocal folds.
Graminoid: Of or relating to grasses, sedges or rushes.
Granular Scales: Small, smooth, round scales that don't overlap.
Gravid: The state of a female carrying eggs.
Gular Fold: A fold of skin across the posterior throat area of some lizards and salamanders.
Hedonic Gland: Any of several glands of various salamanders and reptiles that produce a secretion believed to function in sexual attraction and stimulation.
Hemipenes: Plural of Hemipenis.
Hemipenis: One of a pair of copulatory organs (hemipenes) of snakes and lizards.
Herpetofauna: A collection of the reptile and amphibian species in a given area.
Herps: Slang referring to reptiles and amphibians.
Internal Fertilization: The process where a male deposits his sperm directly into the female's body.
Jacobson's Organ: A chemosensory organ located on the roof of the mouth in all snakes and lizards and many salamanders, used to track prey and find mates.
Keeled Scales: Scales with a lengthwise, narrow ridge down their center.
Lateral: Of, at, toward, or from the side or sides.
Loreal Pits: Region located between the eye and nostril of a snake, functions as infrared detection organ in rattlesnakes.
Melanophore: A pigment-containing cell, present in some reptiles and amphibians that can disperse or contract the pigment within the cell wall, causing an animal's overall coloration to be lighter or darker.
Mental Glands: The mental gland is an oval-shaped pad which develops exclusively in reproductive male salamanders.

Nasolabial Groove: A very small groove extending from each nostril to the upper lip in the Plethodontidae (Lungless Salamanders).
Nubbin: A small lump or residual part, as in bones or cartilage.
Ocular Auto-hemorrhaging: An anti-predation adaptation involving special musculature that constricts to squirt an aimed stream of blood from the eye, performed by certain species within the Phrynosoma (Horned Lizard) genus.
Oligotrophic: Especially of a lake, relatively low in plant nutrients and containing abundant oxygen in the deeper parts.
Overstory: The highest layer of vegetation in a forest, usually forming a canopy.
Oviparous: Egg-laying animal.
Ovoviviparous: Describes animals that retain eggs with shells or membranous coatings that hatch within the female's oviduct, after which the young are born alive.
Parietal Eye: A "third eye" located on top of the head of some lizards that acts as a radiation dosimeter.
Plastron: Underside of a turtle's shell.
Plethodontid: A term referring to members of the Lungless Salamanders (Plethodontidae).
Post-Anal Scales: A scale positioned below the cloaca, at the base of the tail in snakes and lizards. This feature is paired and enlarged in male Iguanids.
Prehensile Tail: Special musculature in certain animals' tails, allowing them to hang from and climb with their tail.
Preoculars: The area in front of the eyes.
Rostral: Situated or occurring near the front end of the body, especially in the region of the nose and mouth.
Scutes: A thickened horny or bony plate on a turtle's shell or the large ventral scales of some snakes and lizards.
Sexual Dimorphism: When males and females of the same species have distinguishable features from one another (coloration, size, shape, etc.).
Snout-Vent Length: Measurement taken from the tip of the head to the cloaca.

Thermoregulation: Regulation and maintenance of core body heat within a certain temperature range.
Understory: The layer of vegetation beneath the main canopy in a forest.
Venter: The belly or underside of an animal.
Ventral Side: Of, on, or relating to the underside of an animal or plant; abdominal area or belly.
Viviparous: Live-bearing animal
Waterdog: The common name for the aquatic larva of the Tiger Salamander (Ambystoma tigrinum), frequently used by anglers in bass fishing.

California Legless Lizard burrowing into loamy soil at the base of an adjacent Bush Lupine.

Salamander Features

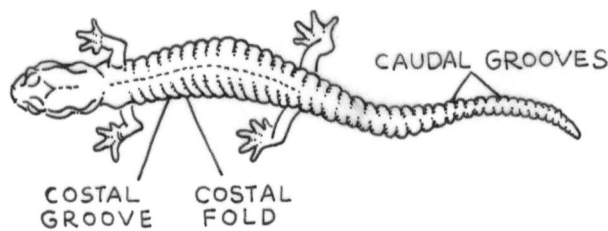

Prominent features of salamanders. Illustration by Sasha Taus.

All members of the Plethodontidae family contain a nasolabial groove. Illustration by Sasha Taus.

Scale Types (Lizards and Snakes)

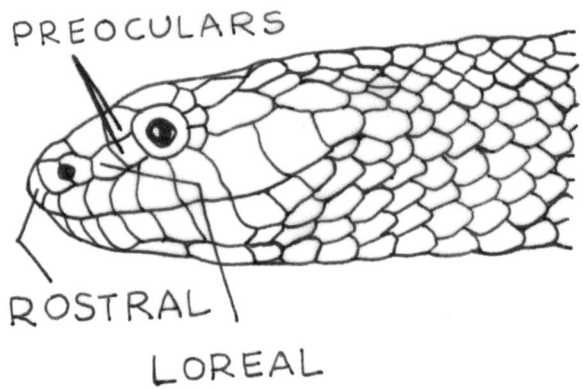

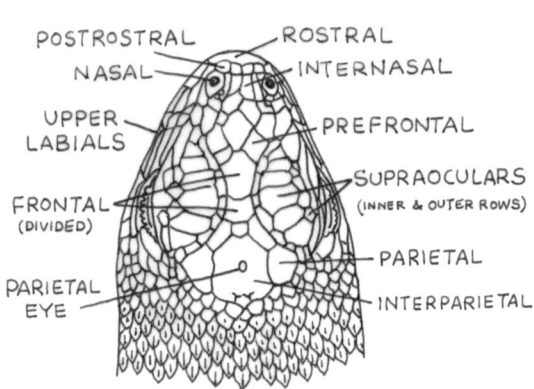

Lizard and Snake scale types. Illustrations by Sasha Taus.

RATTLE

Key feature at the end of the tail in rattlesnakes. Illustration by Sasha Taus.

Baby Rattlesnake in a defensive pose.

Works Cited

Bhattacharjee, Payel, and Debasish Bhattacharyya. "Factor V activator from Daboia russelli russelli venom destabilizes β-amyloid aggregate, the hallmark of Alzheimer disease." Journal of Biological Chemistry 288.42 (2013): 30559-30570.

Bikle, Anne. Amphibians of the Landels-Hill Big Creek Reserve. Environmental Field Program, University of California, Santa Cruz, 1985.

Bondi, Cheryl A., and Sharyn B. Marks. "Differences in flow regime influence the seasonal migrations, body size, and body condition of western pond turtles (Actinemys marmorata) that inhabit perennial and intermittent riverine sites in northern California." Copeia 2013.1 (2013): 142-153.

California Herps: A Guide to the Amphibians and Reptiles of California, edited by Gary Nafis.

Conant, Roger, and Joseph T. Collins. A field guide to reptiles & amphibians: eastern and central North America. Vol. 12. Houghton Mifflin Harcourt, 1998.

Cooper, W. E., and L. J. Vitt. 1985. Blue tails and autotomy: enhancement of predation avoidance in juvenile skinks. Z. Tierpsychol. 70: 265-276.

Frick, Winifred F. "A Field Guide to the Lizards of the Granite Mountains." University of California, Santa Cruz, 1997.

Jennings, Mark R. "Impact of the Curio Trade for San Diego Horned Lizards (Phrynosoma Coronatum Blainvillii) in the Los Angeles Basin, California: 1885-1930." Journal of Herpetology, vol. 21, no. 4, 1987, pp. 356–358. JSTOR, www.jstor.org/stable/1563985.

Jones, Lawrence LC, and Robert E. Lovich. Lizards of the American Southwest: a photographic field guide. Rio Nuevo Publishers, 2009.

Kissner, Kelley J., et al. "Sexual Dimorphism in Malodorousness of Musk Secretions of Snakes." Journal of Herpetology, vol. 34, no. 3, 2000, pp. 491–493. JSTOR, www.jstor.org/stable/1565381.

Kumar, Anoop, Phillip B. Gates, and Jeremy P. Brockes. "Positional identity of adult stem cells in salamander limb regeneration." Comptes rendus biologies 330.6 (2007): 485-490.

Lane, Robert S., and G. B. Quistad. "Borreliacidal factor in the blood of the western fence lizard (Sceloporus occidentalis)." The Journal of Parasitology (1998): 29-34.

Lane, Robert S., and Jenella E. Loye. "Lyme disease in California: interrelationship of Ixodes pacificus (Acari: Ixodidae), the western fence lizard (Sceloporus occidentalis), and Borrelia burgdorferi." Journal of Medical Entomology 26.4 (1989): 272-278.

Middendorf III, G.A.; Sherbrooke, W.C.; Braun, E.J. (2001). "Comparison of Blood Squirted from the Circumorbital Sinus and Systemic Blood in a Horned Lizard, Phrynosoma cornutum". The Southwestern Naturalist.

Pickwell, Gayle Benjamin. Amphibians and reptiles of the Pacific states. Stanford University Press, 1949.

Riley, Seth PD, et al. "Hybridization between a rare, native tiger salamander (Ambystoma californiense) and its introduced congener." Ecological Applications 13.5 (2003): 1263-1275.

Roach, John. "Fear of Snakes, Spiders Rooted in Evolution, Study Finds." National Geographic News (2001).

Ruse, Michael, and Lynne A. Isbell. "The Fruit, the Tree, and the Serpent: Why We See So Well." (2012): 175-177.

Stebbins, Robert C., and Samuel M. McGinnis. Field Guide to Amphibians and Reptiles of California: Revised Edition. University of California Press, 2012. JSTOR, www.jstor.org/stable/10.1525/j.ctt1pn65t.

Stebbins, Robert Cyril. A field guide to western reptiles and amphibians. Houghton Mifflin Harcourt, 2003.

University Of California - San Diego. (2002, March 5). Proliferation Of Argentine Ants In California Linked To Decline In Coastal Horned Lizards. ScienceDaily. Retrieved March 8, 2017 from www.sciencedaily.com/releases/2002/02/020227071151.html

Zeiner, David C., William F. Laudenslayer, and Kenneth E. Mayer. California's Wildlife: Amphibians and reptiles. Vol. 1. State of California, the Resources Agency, Department of Fish and Game, 1988.

Zweifel Richard, G. "Cogger, HG & Zweifel, RG ed. Encyclopedia of Reptiles and Amphibians." (1998).

Pair of Southern Alligator Lizards found copulating underneath a coverboard.

Index

Actinemys marmorata 17, 63, 64
Alligator Lizard 31-32
    Northern 35-37, 99
    Southern 33-34, 97
Ambystoma californiense 14, 82-85
Aneides lugubris 14, 78-81
Anniella pulchra 11, 16, 38-40
Anti-Predation Adaptations 10, 11
Arboreal Salamander 14, 78-81
Autotomy 10, 11, 70, 74
Batrachoseps gavilanensis 14, 74-77
California Tiger Salamander 14, 82-85
Coast
    Garter Snake 15, 45-47
    Horned Lizard 7, 11, 16, 27-30
Coluber constrictor mormon 48-51, 59
Conservation Note 12
Contia tenuis 15, 55-57
Crotalus oreganus oreganus 41-44, 61, 62, 94
Diadophis punctatus vandenburghii 15, 52-54
Dichotomous Key 13
Elgaria 31-32
    coerulea 35-37, 99
    multicarinata 33-34, 97
Ensatina eschscholtzii eschscholtzii 14, 70-73
Foraging Strategies 8
Frog 17, 65-68
Gabilan Slender Salamander 14, 74-77
Lampropeltis getula 15
Lizard 16, 19-40
Methods 4
Monterey
    Ensatina 14, 70-73
    Ring-necked Snake 15, 52-54
Northern California Legless Lizard 11, 16, 38-40
Pacific
    Chorus Frog 17, 65-68
    Gopher Snake 15, 44, 51, 58-60
    Pacific Rattlesnake 15, 41-44, 61, 62, 94

Phrynosoma blainvilli 7, 11, 16, 27-30
Pituophis catenifer catenifer 15, 44, 51, 58-60
Plestiodon skiltonianus skiltonianus 2, 16, 23-26
Plethodontidae 69-81
Preface 2
Pseudacris regilla 17, 65-68
Reproduction 9
Salamander 14, 69-85
    Features 80, 92
Scale Types (lizards and snakes) 93
Sceloporus occidentalis 16, 19-22
Snake 15, 41-62
Sharp-tailed Snake 15, 55-57
Skilton's Skink 2, 16, 23-26
Thermoregulation 7
Thamnophis elegans terrestris 15, 45-47
Turtle 17, 63, 64
Western
    Fence Lizard 16, 19-22
    Pond Turtle 17, 63, 64
    Yellow-bellied Racer 15, 48-51, 59
Works Cited 95-97

About the Author

The author holding a Northern Alligator Lizard.

Max Taus is an avid herper and naturalist studying Environmental Studies and Molecular, Cell and Developmental Biology at the University of California, Santa Cruz. When not hunting for herps, he passionately pursues music, mountain biking and glassblowing. He aspires to one day become a full-fledged field biologist and spread his passion and knowledge of the natural world, in hopes of inspiring a lasting joy for the great outdoors in others!

Contact information: maxlevitaus@gmail.com

www.ingramcontent.com/pod-product-compliance
Lightning Source LLC
Chambersburg PA
CBHW040316220526
45473CB00009B/2458